Power Interconnection in Southeast Asia

Providing an analysis of multilateral power markets, this book examines power interconnection in Southeast Asia, especially among the ASEAN countries.

It uses evolutionary experience of electricity interconnection and trade in three international markets that have relevance for ASEAN to draw upon common global themes. Specifically, it compares the Southern African Power Pool, the European grid, and Nord Pool. Discussing the progress made among ASEAN countries in regional energy integration, with a particular focus on the Greater Mekong Sub-region interconnection, it also examines the recently announced interconnection concept between Lao People's Democratic Republic, Thailand, Malaysia, and Singapore.

Exploring the challenges facing ASEAN interconnection of power grids in the context of previous experience elsewhere in the world, this book presents a template for appropriate best practice in terms of technical, political, and financial requirements. It will therefore be of value to decision makers interested in the political economy of energy in Southeast Asia, as well as academics working on Energy Politics and Southeast Asian Politics.

Anthony David Owen is Principal Fellow and Head of Energy Economics at the Energy Studies Institute (ESI), National University of Singapore.

Anton Finenko is currently Sales Operations Manager at Siemens Gas and Power, Singapore.

Jacqueline Tao is Research Associate at Wood Mackenzie, Singapore.

Routledge Contemporary Southeast Asia Series

The aim of this series is to publish original, high-quality work by both new and established scholars on all aspects of Southeast Asia.

The Appeal of the Philippines
Spain, Cultural Representation and Politics
José Miguel Díaz Rodríguez

Propaganda and the Genocide in Indonesia
Saskia E. Wieringa with Nursyahbani Katjasungkana

The International People's Tribunal for 1965 and the Indonesian Genocide
Edited by Saskia E. Wieringa, Jess Melvin and Annie Pohlman

Singapore's Multiculturalism
Evolving Diversity
Chan Heng Chee, Sharon Siddique with Irna Nurlina Masron, Dominic Cooray

Researching China in Southeast Asia
Edited by Ngeow Chow Bing

Religious Actors and Conflict Transformation in South East Asia
Indonesia and the Philippines
Edited by Jürgen Rülend, Christian von Lübke and Marcel M. Baumann

Power Interconnection in Southeast Asia
Anthony David Owen, Anton Finenko, and Jacqueline Tao

For more information about this series, please visit: www.routledge.com/Routledge-Contemporary-Southeast-Asia-Series/book-series/RCSEA

Power Interconnection in Southeast Asia

Anthony David Owen,
Anton Finenko, and
Jacqueline Tao

Routledge
Taylor & Francis Group
LONDON AND NEW YORK

First published 2019
by Routledge
2 Park Square, Milton Park, Abingdon, Oxon OX14 4RN

and by Routledge
52 Vanderbilt Avenue, New York, NY 10017

Routledge is an imprint of the Taylor & Francis Group, an informa business

British Library Cataloguing-in-Publication Data
A catalogue record for this book is available from the British Library

Library of Congress Cataloging-in-Publication Data
Names: Owen, Anthony David, author. | Finenko, Anton, author. | Tao, Jacqueline, author.
Title: Power interconnection in Southeast Asia / Anthony David Owen, Anton Finenko and Jacqueline Tao.
Description: New York : Routledge, 2019. | Series: Routledge contemporary Southeast Asia series | Includes bibliographical references and index.
Identifiers: LCCN 2018059895 (print) | LCCN 2019000055 (ebook) | ISBN 9780429424526 (Ebook) | ISBN 9780429755262 (Adobe Reader) | ISBN 9780429755255 (ePub) | ISBN 9780429755248 (Mobipocket Encrypted) | ISBN 9781138388567 | ISBN 9781138388567 (hardback) | ISBN 9780429424526 (ebook)
Subjects: LCSH: International interconnected electric utility systems—Southeast Asia. | Electric power distribution—Southeast Asia—International cooperation. | ASEAN.
Classification: LCC HD9685.A7852 (ebook) | LCC HD9685.A7852 O94 2019 (print) | DDC 333.793/20959—dc23
LC record available at https://lccn.loc.gov/2018059895

ISBN: 978-1-138-38856-7 (hbk)
ISBN: 978-0-429-42452-6 (ebk)

Typeset in Times New Roman
by Apex CoVantage, LLC

Contents

Figures

Tables

Acknowledgements

This study was undertaken by the Energy Studies Institute (ESI), National University of Singapore, as a part of a research project funded by the Konrad-Adenauer-Stiftung. Many experts from outside of the ESI provided valuable insights that became part of this document. In particular, the authors wish to thank Joseph Cheng, Sakshi Balani, and Ashwin Dhandhania for their assistance in the preparation of the research report that eventually lead to this book.

The authors also wish to thank Peter Hefele, Director, Regional Project, Energy Security and Climate Change Asia-Pacific (RECAP), Konrad-Adenauer-Stiftung, for his support and encouragement during the research project. The opinions expressed in this article are those of the authors and do not reflect the views of the Konrad-Adenauer-Stiftung.

The authors gratefully acknowledge permission to reproduce diagrams and/or tables from the following sources: Association of Southeast Asian Nations (ASEAN), ESI, Heads of ASEAN Power Utilities/Authorities (HAPUA), International Energy Agency (IEA), International Renewable Energy Agency (IRENA), Professor Rudi A. Hakvoort, the World Bank, and the Southern African Power Pool (SAPP).

Abbreviations and acronyms

ACE	ASEAN Centre for Energy
ACER	Agency for the Cooperation of Energy Regulators
ADB	Asian Development Bank
AEC	ASEAN Economic Community
AEMI	ASEAN Energy Market Integration
AERN	ASEAN Energy Regulators' Network
AIMS	ASEAN Interconnection Master Plan Studies
APAEC	ASEAN Plan of Action for Energy Cooperation
APG	ASEAN Power Grid
APGCC	ASEAN Power Grid Consultative Committee
ASEAN	Association of South-East Asian Countries
BBL	Billion Barrels
CBA	Cost-Benefit Analysis
CEC	Copperbelt Energy Corporation
CEE	Central and Eastern Europe
CEER	Council of European Energy Regulators
CfD	Contract for Difference
CHP	Combined Heat and Power
CO_2	Carbon Dioxide
COD	Commercial Operations Date
COMESA	Common Market for Eastern and Southern Africa
CSG	China Southern Power Grid
CWE	Central and Western Europe
DIS	Directorate of Infrastructure and Services
DRC	Democratic Republic of the Congo
DSOs	Distribution System Operators
EC	European Commission
EDC	Electricité du Cambodge
EDL	Electricité du Laos
EDL-Gen	EDL-Generation

EDM	Electricidade de Mozambique
EE	Economic Exchange
EGAT	Electricity Generating Authority of Thailand
EIA	Energy Information Administration
EMA	Energy Market Authority, Singapore
EMC	Energy Market Company
ENTSO-E	European Network of Transmission System Operators for Electricity
EPAD	Electricity Price Area Differentials
EPPO	Energy Policy and Planning Office
ERC	Energy Regulatory Commission
ERIA	Economic Research Institute for ASEAN and East Asia
ERRA	Energy Regulators Regional Association
ESI	Energy Studies Institute, Singapore
ETS	Emissions Trading Scheme
EU	European Union
EU-ETS	European Union Emission Trading Scheme
FPM	Forward Physical Market
FPM-M	Forward Physical Market – Monthly
FPM-W	Forward Physical Market – Weekly
GDP	Gross Domestic Product
GHG	Greenhouse Gas
GMS	Greater Mekong Sub-region
Gt	Gigaton
GW	Gigawatts
GWh	Gigawatt-hour
HAPUA	Heads of ASEAN Power Utilities/Authorities
HPP	Hydropower Plant
HVAC	High-voltage Alternating Current
Hz	Hertz
ICEM	International Centre for Environmental Management
ICER	International Confederation of Energy Regulators
IEA	International Energy Agency
IEEJ	Institute of Energy Economics, Japan
IGSO	Independent Grid System Operator
IPP	Independent Power Producer
IRENA	International Renewable Energy Agency
IREP	Institute of Renewable Energy Promotion
ISO	Independent System Operator
ITCs	Independent Transmission Companies
ITO	Independent Transmission Operator
KAS	Konrad-Adenauer-Stiftung

Kton	Kiloton
kV	Kilovolt
kW	Kilowatt
kWh	Kilowatt-hour
Lao	PDR Lao People's Democratic Republic
LEC	Lesotho Electricity Corporation
LHSE	Lao Holding State Enterprise
LMB	Lower Mekong Basin
LNG	Liquefied Natural Gas
LPG	Liquefied Petroleum Gas
LTMS	Lao PDR, Thailand, Malaysia, and Singapore
MD	Maximum Demand
MEA	Municipal Electricity Agency
MEDREG	Association of Mediterranean Energy Regulators
MEM	Ministry of Energy and Mines
MFO	Medium Fuel Oil
mmBtu	Million British Thermal Units
MMT	Million Metric Ton
MoE	Ministry of Energy
MoF	Ministry of Finance
MOTRACO	Mozambique Transmission Company
MoU	Memorandum of Understanding
Mt	Million Tonnes
MTI	Ministry of Trade and Industry, Singapore
Mtoe	Million Tonnes Oil Equivalent
MW	Megawatt
MWh	Megawatt-hour
NARUC	National Association of Regulatory Utility Commissioners
NordREG	Nordic Energy Regulator
NP	Non-operating Members
NRAs	National Regulatory Agencies
O&M	Operation and Maintenance
OP	Operating Members
OTC	Over the Counter
PCI	Projects of Common Interest
PEA	Provincial Electricity Agency
PIDA	Programme for Infrastructure Development in Africa
PP	Power Purchase
PPA	Power Purchase Agreement
PRC	People's Republic of China
PV	Photovoltaic
RECS	Regional Economic Communities

REMIT	Regulation on Energy Market Integrity and Transparency
RERA	Regional Electricity Regulators Association
RM	Malaysian Ringgit
RPCC	Regional Power Coordinating Center
RPTCC	Regional Power Trade and Coordinating Committee
SADC	Southern African Development Community
SADCC	Southern African Development Coordination Conference
SAPP	Southern African Power Pool
SB	Single Buyer
SEC	Swaziland Electricity Company
SNEL	Societe Nationale d'Electricite
SPP	Small Power Producer
SPV	Special-Purpose Vehicle
SRMC	Short-Run Marginal Cost
TAGP	Trans-ASEAN Gas Pipeline
TAU	Technical and Administrative Unit
TBA	To Be Advised
TBC	To Be Confirmed
tcf	Trillion Cubic Feet
TNB	Tenaga Nasional Berhad
toe	Tonnes of Oil Equivalent
TPA	Third-party Access
TSOs	Transmission System Operators
TWh	Terawatt-hour
TYNDPs	Ten-Year-Network-Development-Plans
UK	United Kingdom
UN	United Nations
UNSCR	United Nations Security Council Resolution
US$	US Dollar
USAID	United States Agency for International Development
USc	US Cents
VRE	Variable Renewable Energy
VSPP	Very Small Power Producer
XBID	Cross-Border Intraday Market Project
ZESA	Zimbabwe Electricity Supply Authority
ZESCO	Zesco Limited

Introduction

The demand for electricity is steadily increasing in Southeast Asian countries as the result of ongoing economic and population growth. Although these countries have abundant and diversified energy resources, the uneven distribution of these resources and different stages of economic development between individual countries make addressing growing demand requirements challenging. The International Energy Agency (IEA) estimates that the Association of Southeast Asian Nations (ASEAN) will need to deploy, in aggregate, 354 gigawatts (GW) of additional capacity for power generation by 2040 which would require investments of US$618 billion in generation and US$690 billion in the transmission and distribution sectors.[1]

The ASEAN Plan of Action for Energy Cooperation (APAEC) recognizes that integrating power grids and enabling cross-border electricity trade can help ASEAN member states address their regional energy challenges. Interconnections create positive impacts in terms of ensuring energy security and power system reliability, while also creating economic opportunities and the formation of new partnerships. Specifically, there is growing interest for the potential of utilizing hydropower resources in the Lower Mekong region to not only power countries in the region, but also to supplement increasing electricity demand requirements in neighbouring Thailand, Malaysia, and Singapore.

Drawing upon some common themes across three international examples of interconnected grids, we draw lessons for pursuing integration activities between Lao PDR, Thailand, Malaysia, and Singapore (LTMS) and discuss possible approaches for establishing a wider common market for electricity in ASEAN.

The book analyses the development of power interconnection in Southeast Asia and, more specifically, ASEAN, focusing on the proposed interconnection between LTMS. In our analysis, we provide an overview of the evolution of electricity interconnection and trade in three international markets we believe to be relevant for ASEAN, specifically, Southern African

Power Pool (Chapter 2), European electricity markets (Chapter 3), and Nord Pool (Chapter 4). We also discuss the progress among ASEAN countries in regional energy integration with a particular focus on the Greater Mekong Sub-region (GMS) interconnection, the first significant project in ASEAN that involves several countries' trading power.

In the review of the experience of selected regional electricity markets around the world, we identify some key elements of integration that emerged independently as those markets have evolved. These are as follows:

- Coordinated physical infrastructure development;
- Standardized and harmonized rules of operation;
- Some form of market competition; and
- Empowered governing or coordinating institutions.

Prioritization of these elements and the sequence of steps to achieve them are not straightforward, as they depend on current circumstances of the regional market's environment and history. As such, these elements are still undergoing development in the international markets under review in this book.

Market integration in Europe adopted a top-down integration approach, capitalizing on the legal system of the European Union (EU) and mature country power networks. In contrast, Nordic and Southern African markets developed on an incremental and voluntary basis, driven by the utilities themselves: a bottom-up approach. Given diverse regional circumstances in ASEAN, and the absence of an overarching legal system like in the EU, we believe that the latter approach is more suitable for ASEAN than the EU model.

The importance of coordinated infrastructure development is particularly important in markets with growing electricity demand, such as the Southern African Power Pool (SAPP) and ASEAN. Insufficient generation and transmission infrastructure in Southern Africa seriously limit the progress of the otherwise successfully evolving market and undermine the benefits of market integration. Lack of infrastructure development there is driven by non-cost-reflective tariffs, low market transparency, and weak protection of third-party investors. These aspects deserve consideration by ASEAN, where the required generation capacity is expected to double by 2040.

Another important question is whether deregulated electricity sectors in particular ASEAN countries create barriers for cross-border power trade. While this is a matter of national policy in each sovereign country, we note that deregulated markets facilitate cross-border power trade only at a fairly advanced stage and thus are not a limiting factor during the early stages. Nevertheless, separation of generation and transmission is highly recommended in order to increase efficiency in electricity markets. Choosing an appropriate market design is particularly challenging for the LTMS project, where Lao PDR, Thailand, and Malaysia have single-buyer models, while Singapore has a competitive electricity market. Hydropower imports may also undermine the competitiveness of the non-subsidized gas-fired power generation sector in Singapore.

Tapping on hydropower potential in the Mekong Basin is a crucial aspect in the current vision of ASEAN electricity market integration. We note that virtually all existing regional interconnection studies do not sufficiently analyse the profound environmental and socioeconomic impacts caused by the planned damming of the Mekong and its tributaries. These impacts are loss of biodiversity of global importance, increased food insecurity for millions of people, and increased international tensions – all of which could outweigh the collaborative push resulting from electricity market integration. All these aspects should be considered by policymakers in ASEAN when designing a vision of a common resource use.

Given the commitment of ASEAN member countries to increase cross-border interconnection and power trade, we conclude the book with three market design options and detail the required steps to achieve them. In setting out these options, we have sought to incorporate important lessons derived from international experience analysed in this study.

Note

1 The terms "Southeast Asia" and "ASEAN" are often treated as synonyms. ASEAN membership consists of the following sovereign states (in alphabetical order): Brunei Darussalam, Cambodia, Indonesia, Laos PDR, Malaysia, Myanmar, the Philippines, Singapore, Thailand, and Vietnam. Southeast Asia is generally regarded as these ten ASEAN nations plus Timor Leste. The latter has observer status with ASEAN. Papua New Guinea also has observer status, although it is generally classified as being in Oceania rather than Southeast Asia (but not without some controversy!). In this text, we will focus on ASEAN and its member states, and therefore only use the term Southeast Asia when referring to the larger grouping of nations.

1 The political economy of electricity interconnections

Introduction

An electricity interconnector is a physical link which permits the transfer of electricity across two separate control areas (e.g. state or international boundaries), each with their own separate market operators. Interconnector companies derive their revenue from arbitrage arising from price differentials between the two areas.[1] Thus, interconnectors should ensure that wholesale electricity prices are equalized across the boundaries unless a physical or political constraint is present, or a profit maximizing merchant interconnector indulges in some form of strategic behaviour.

In general, the welfare benefits from interconnection are broadly similar to those that arise from international trade. However, even if there is an overall net improvement in welfare across jurisdictions, its distribution may be skewed to the extent that it creates both winners and losers. In addition, positive market externalities (e.g. enhanced security of supply) and unpriced environmental impacts (e.g. reduction in greenhouse gas (GHG) emissions) may be present that private investors cannot capture. The latter, therefore, may be reluctant to invest in interconnectors, even though, in a cost-benefit framework, they provide net welfare benefits to society as a whole.

This chapter provides an overview of the theoretical background to the discussion on interconnections, highlighting the associated benefits and costs, as well as some factors affecting the quantification of these benefits and costs.

Welfare effects of interconnection

In order to evaluate the net economic value to society of an interconnection, a cost-benefit framework can be utilized. The costs are predominantly made up of the initial capital investment (which will be known if the interconnector has already been constructed), plus future operation and maintenance costs over the investment's planned lifespan, which will likely constitute only a small part of the total.

The benefits, however, are far less well-defined and hard to derive due to challenges in attaching monetary values for some benefits. In addition, there is no guarantee that positive welfare benefits will actually arise from such an investment.

For a new connector, benefits may arise in the following ways (adapted from Turvey 2006):

- Postponing investments in new generation capacities to meet domestic demand;
- Savings in terms of reductions in unserved energy, typically measured in terms of value of lost load;
- Fuel and operating costs savings (net of transmission and distribution losses) due to the substitution of costly generation sources by cheaper ones;
- Cost savings from reduction in frequency control, spinning reserves, and other ancillary service costs;
- Strengthening of energy security;
- Increasing competition in energy supply, and
- Climate change mitigation through exploitation of sustainable energy sources.

Secondary, or indirect, economics benefits may also arise due to increased power system stability and reliability. Local economic and infrastructure development, improvements in public infrastructural services, as well as the "re-spending" effect whereby savings as a result of cheaper electricity prices translate into more disposable income may all arise due to improved power supply reliability. Moreover, interconnection allows for market expansion into neighbouring markets, which may incentivize investments into power generation facilities, contributing to lower electricity prices in the long run.

The occurrence and extent of all these benefits may differ significantly according as to whether the markets being interconnected have different generation technologies and demand profiles.

Quantifying the benefits of interconnectors

Real welfare effects may arise through increases in both productive and allocative efficiency. The former can occur when cheaper generation sources replace more expensive ones as a result of the interconnector. Thus, if Country A has a lower marginal cost of electricity production than Country B, it would be welfare enhancing in aggregate to generate the marginal power in Country A rather than Country B.

Country A could export electricity but would also need to generate more to satisfy the resulting increase in total demand, leading to a move up the merit order and consequently higher wholesale prices and higher infra-marginal rent. Thus, generators in Country A will benefit, while domestic consumers will suffer, from higher prices. The reverse will happen in Country B. Imports will displace generators in the merit order, resulting in lower prices and lower infra-marginal rent for generators, but the benefit of lower prices for consumers.

How such trade can increase aggregate welfare is illustrated, in a static context, in Figure 1.1, where the line NS_A represents net supply in the exporting country, A. This line shows how the price in Country A would increase as a function of export capacity. Similarly, ND_B represents net demand in the importing country, B, showing a price decrease in the country when importing a given capacity.

K_0 represents the situation where there is no interconnector capacity between the two countries. Assume that Country A now has a transmission capacity to export of K_1, then the price in Country A would rise to P_A, while the price in the importing Country B would fall to P_B. The triangle, BCD,

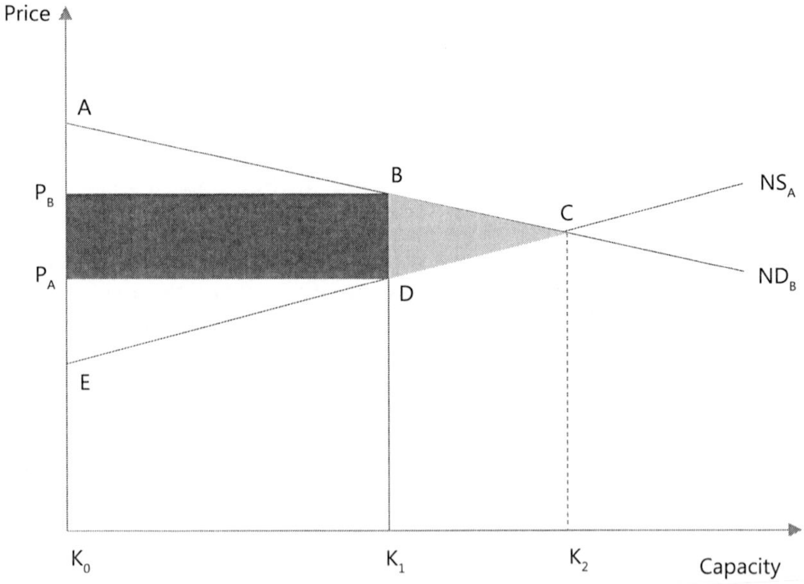

Figure 1.1 Interconnection Capacity between Country A and B

Source: Adapted from Hakvoort and de Jong (2007)

reflects the deadweight loss arising from a lack of sufficient transmission capacity. From the perspective of a private investor, the revenue benefits of the interconnector are given by the area P_AP_BBD (i.e. the marginal price of congestion, P_B-P_A, multiplied by the capacity of the new interconnector, K_1-K_0). This is the area that a merchant interconnector will attempt to maximize. From a social perspective, however, the value of the interconnector is given by the area of the trapezium ABDE. The actual distribution of this social value between the consumer and producer surplus depends on who actually pays for using the interconnector and who then benefits from the resulting revenues.

Note that if there were no constraints – i.e. the interconnector capacity is K_2 – the social value of the interconnector is maximized at this point and given by the triangle ACE. However, from a private investor perspective, the revenue benefits are zero at this point. Thus, it is in the financial interests of a merchant interconnector to have a degree of congestion in the system; otherwise, there would be no revenue from the investment to cover its capital costs. By implication, a regulated interconnector would not need to maintain a price differential since it could recover its capital costs from its regulated assets base. However, if it is assumed that (P_B-P_A) reflects the unit cost of capacity (i.e. we are assuming that the interconnector is not reaping monopoly profits and that operating costs are negligible), this could actually result in a net welfare loss compared with the constrained case, since the area BCD is smaller than the area given by $(P_B\text{-}P_A)*(K_2\text{-}K_1)$.

Factors affecting the benefits of interconnection

The benefits of interconnection are likely to increase with diversity in the generation technology mix between the two countries. Thus, a country with a high capacity of variable, low short-run marginal cost (SRMC), and technologies (such as wind and solar) may export during favourable weather conditions to the neighbouring country that relies almost exclusively on fossil fuel generators. While this asymmetry may increase the value of the interconnector, the volatility of the variable technologies may result in uncertainty as to the direction of flow on it. In addition, wholesale electricity prices generally track changes in the daily pattern of demand which is driven by consumer behaviour and industrial requirements. Thus, price differences between two countries are likely to be more favourable for interconnectors when their respective demand profiles differ.

Security of supply benefits of interconnectors

The more isolated the market, the higher the level of installed capacity required to meet peak demand and regulating and contingency reserves.

Thus, interconnection with a neighbouring country's electricity market will permit both countries to not only reduce price volatility but also reduce the requirement for peaking and contingency back-up plant that is rarely used. However, some jurisdictions may be uncomfortable with reliance on imports to cover extreme events and would interpret security of supply to mean "domestic supply".

In the case of major interconnections, supply security must involve ample spinning reserve to address the case of an unexpected complete loss of supply through the interconnector.

Environmental costs and benefits

To the extent that an interconnector reduces GHG emissions, the benefits will accrue explicitly in the cost-benefit analysis (CBA) if the countries involved have a carbon market, although the values will reflect control rather than damage costs. Logically, the former should be less than the latter, so one would expect an underestimate of the benefits.

Local environmental impacts are clearly site-specific, most of which would probably have their greatest impact during the construction stage. However, with hydropower, one issue is security in times of drought which implies that sufficient back-up power must be available to take up any supply shortfall. There are also negative issues associated with constructing dams on river ecosystems, which can change the natural flow with resulting adverse impacts on downstream fisheries and agriculture.

For further reading, a very comprehensive review of issues arising from development and use of power interconnections is provided in UN (2006).

Note

1 Merchant interconnectors operate in this fashion. Regulated interconnectors, as the name suggests, operate and receive revenue according to a "regulated rate" imposed by the relevant market operator(s).

References

Hakvoort, R. A., and H. M. de Jong. 2007. "Pushing European Power Transmission: Private Investment in Priority Interconnections?" *European Review of Energy Markets* 4: 109–140.

Turvey, R. 2006. "Interconnector Economics". *Energy Policy* 34: 1457–1472. doi:10.1016/j.enpol.2004.11.009.

UN (United Nations). 2006. *Multi-Dimensional Issues in International Electric Power Grid Interconnections*. Washington, DC: United Nations. Accessible at: https://sustainabledevelopment.un.org/content/documents/interconnections.pdf.

2 The Southern African Power Pool

Introduction

The Southern African Power Pool (SAPP) is the first and probably the most relevant case study for potential power interconnection projects in ASEAN. This multilateral trading scheme was established in 1995 between power utilities of 12 Southern African countries, although bilateral power trade and existing interconnections between Southern African power utilities predated the SAPP. The SAPP has also established a power exchange with day-ahead and intraday spot trading in a move to a more competitive market structure. There are also plans to add an ancillary services market and to further improve the performance of existing markets. Currently, approximately 5% of regional electricity generation is traded through the exchange.

The generation and transmission infrastructure in the SAPP are fairly well-developed, partly dating back to the discovery and rise of the Copperbelt during colonial times. Later, several large hydropower projects were built and connected to the main grid with the help of international donors during the 1960s and 1970s. The interconnectors allowed better sharing of available energy resources in the region and diversification of electricity supply sources. This is particularly important for the northern region with ample hydropower which is exposed to droughts and man-made interruptions to water supply.

Years of ongoing economic development created the need for additional generation and transmission infrastructure in the SAPP. The widening shortage of supply capacity caused load shedding and emergency electricity imports for some countries. There are a multitude of reasons behind lagging infrastructure development such as lack of coordinated development, no clear endorsement of intended projects, no aggregated demand forecasts, excessive bureaucracy, and non-cost-reflective electricity tariffs. Many of these issues are pertinent to the countries in ASEAN today, which face similar socioeconomic and political conditions as countries in Southern Africa.

Historical development

A long history of electricity sector cooperation in Southern Africa preceded the official inauguration of the SAPP in the 1990s. The colonial period played a major role in the establishment of the SAPP. Under colonial rule, the development of most generation and transmission projects in the SAPP were closely tied to the inception and growth of mining and industry in today's South Africa, Zambia, Zimbabwe, and Democratic Republic of the Congo (DRC). The expansion of industrial activities to Botswana, Mozambique, and Namibia has also led to the development of hydropower projects in those countries and the linking of their power grids with their neighbours.

The 1980s

The formal push for establishing a power pool in Southern Africa started in 1980, when the Southern African Development Coordination Conference (SADCC) was formed. The SADCC is the predecessor of today's socio-economic cooperation agency, the Southern African Development Community (SADC). A Technical and Administrative Unit (TAU) consisting of SADC officials was formed in Angola. The TAU conducted regular discussion meetings and roundtables that paved the way for energy integration planning in the region. For example, an electricity sub-committee made up of national utilities of the member countries was inaugurated in 1990 as part of a TAU initiative. This sub-committee not only provided technical advisory expertise to the energy ministers but also facilitated cooperation among them. At that point in time, South Africa, with the largest electricity sector in the region, was not allowed to participate in the SADC because of the apartheid policy of its government.

The 1990s

Two key events in the early 1990s led to the formal establishment of the SAPP. The first event was the abolition of apartheid in South Africa, which removed sanctions on the country's participation in regional cooperation. SADC, which replaced SADCC in 1992, included South Africa as a member state. The increased willingness to cooperate with the new government in South Africa paved the way for multiparty intergovernmental and inter-utility agreements between member countries. Some success stories include the signing of agreements for the Cahora Bassa interconnector (signed by Mozambique, South Africa, and Zimbabwe) and the Marimba interconnector (signed by Botswana, South Africa, and Zimbabwe). The formal signing of such agreements facilitated access to financing that brought about the commissioning of a system of 400 kV transmission lines from Zimbabwe to South Africa via Botswana in 1996 and the subsequent extension to Cahora Bassa (commissioned in 1997).

The second key event was the unprecedented drought of 1991–1992. This extreme weather event restricted large amounts of hydropower production in the Zambezi Basin, contributing to severe load shedding events in Zimbabwe and Zambia. These negative economic and social consequences of load shedding during periods of severe drought exposed the over-reliance of the northern network on the weather-sensitive hydropower and the lack of any drought-resistant back-up. As a consequence, Zimbabwe, Zambia, and the DRC signed a multiparty agreement to source power from the DRC to Zimbabwe, with Zambia being a transit party. Interconnection projects, therefore, provided diversification of supply across markets, allowing the stable baseload coal-fired generators in South Africa to act as back-up during extreme weather conditions.

Inauguration of the Power Pool

By the mid-1990s, the SADC region showcased numerous examples of successful cooperation between national utilities/governments and developed physical interconnection links among SADC members. The Electricity Subcommittee took advantage of these past successes to lay the foundations of the framework agreements underpinning the Power Pool. This paved the way for the SAPP to be created in August 1995. The SAPP was established from a neutral institutional standpoint as a non-binding entity. It was not expected to forcefully intervene in trading arrangements between members and national capacity development plans. Rather, it was intended as an institutional framework enabling a common vision for an interconnected electricity market for the future benefit of the whole region. The formal establishment of the SAPP was also a strengthening point for the region's political agenda of economic integration.

The 2000s

In its first five years, the SAPP relied solely on bilateral and multilateral agreements between the member countries. In 2001, the SAPP short-term spot market, which was designed as a precursor to a future fully competitive electricity market, was established. However, in 2007, this market was closed down due to limited power supply and transmission capacity.

In 2006, the SAPP received a grant from the government of Norway and the Swedish International Development Cooperation Agency to develop a spot exchange with day-ahead and intraday markets. The day-ahead market was developed by 2007, but its implementation was delayed due to an ongoing study on nodal pricing for transmission wheeling and losses. After the testing phase between 2007 and 2009, the day-ahead market was launched in late 2009. The intraday market was subsequently inaugurated in 2013.

Benefits of market integration

It should be noted that no centralized study on feasibility – i.e. costs and benefits – had been done for the SAPP before it was formally launched. However, participating utilities performed their own reviews of existing bilateral and multilateral electricity interconnection schemes and exchanged visits with experts from the established power pools in North America and Europe before launching the SAPP. After its establishment, studies were performed in crucial areas with the support of international donors. In 2002, a cost-benefit study funded by the United States Agency for International Development (USAID) concluded that there were both immediate and long-term benefits to be reaped through the coordination of electricity investments. An estimated benefit of US$100 million per annum could be gained through short-run optimization of regional opportunities and a total of US$1.5 billion in net present value of investments could be saved in the long term. Further, the coordination of regional capacity development plans under one umbrella could result in savings of US$48 billion by 2025, compared to a scenario with uncoordinated capacity plans undertaken by each country.

Vision and objectives

The SAPP was formally launched with the vision and objectives of developing a competitive electricity market while adhering to the political principle of regional economic integration. SAPP's vision is an integrated electricity market that will (SAPP 2016)

- facilitate the development of a competitive electricity market in the SADC region;
- give the end user a choice of electricity supplier;
- ensure that the Southern African region is the region of choice for investment by energy intensive users; and
- ensure sustainable energy developments through sound economic, environmental, and social practices.

The SAPP was launched as a "cooperative pool" scheme which advocated that SAPP members should share information about generation and transmission costs for price setting purposes. Based on similar principles of pre-existing power exchange agreements, the prevailing price is set as the midpoint between the marginal or average cost of generation and the avoided costs for the importer. Moving forward, the SAPP would have to graduate to a "competitive pool" where prices would be market-determined (i.e. on the basis of a supply-demand equilibrium). Liberalization of electricity markets has been a major trend around the world since the 1990s. Liberalized markets not only allow participants more freedom in choosing their electricity suppliers but also introduce some degree of market flexibility, which allows players to respond to short-term, supply-demand fluctuations. It is also theoretically beneficial in promoting private sector participation, albeit that there is no unanimous expert view on this issue.

Despite its advocacy for expanding electricity access in rural communities, the SAPP efforts have been largely concentrated on large-scale transmission interconnections, perhaps due to its industry-driven history. This is in contrast to other regional power organizations, such as the West African Power Pool, which engages in various cross-border distribution and interconnections projects that focus on rural electricity access. However, there is documented evidence that rural electrification is a major part of the SAPP scheme (SAPP 2016).

Geographical coverage

The SAPP can generally be divided into two networks, shown in Figure 2.1, which rely on different electricity supply sources. The southern region has mainly coal-fired thermal capacity, while the northern region mainly uses hydropower. The interconnection allows the southern part of the region to take advantage of cheap hydropower generated in the north during the rainy season, while the net flow is reversed during the dry season when the thermal south could export energy to its northern neighbours.

Southern region

The southern region comprises South Africa, Lesotho, Swaziland, Botswana, and Namibia. Of these countries, South Africa has historically been the dominant player, both as a producer and as a consumer. South Africa, being a large industrialized nation, developed an extensive coal-fired generation capacity that exported excess electricity to its neighbours. Furthermore, the national grids of Botswana, Lesotho, Swaziland, and Namibia were developed as extensions of the South African grid. The Cahora Bassa hydroelectric project was developed in 1979 to supply hydropower from Mozambique to South Africa.

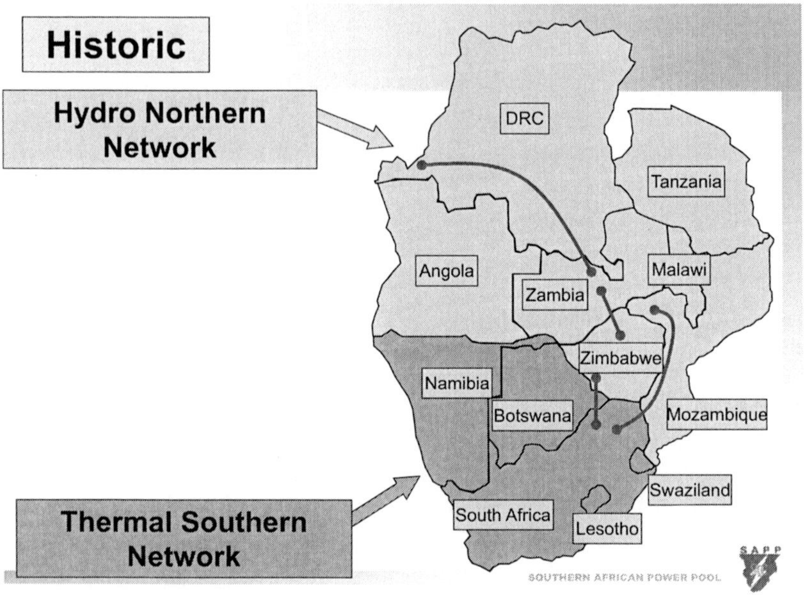

Figure 2.1 Northern and Southern Regions of the SAPP

Source: Beta (2013)

Northern region

The northern region has a large number of hydropower projects located on the Zambezi and Congo rivers. These projects were developed to address the growing electricity demand of the mining industry in the Copperbelt. One of the most notable grid interconnection projects was the Kariba scheme in the northern area, which was developed in the 1950s to supply electricity to the Copperbelt and runs through (what is now) northern Zambia and the southern DRC. Since Zambia's independence in 1964, various donor agencies have contributed to building the 900 MW Kafue Gorge power station and the 600 MW Kariba North power station to support the effort towards Zambian energy self-sufficiency.

Through the use of an interconnector built in with Zambia in 1956, the Inga hydroelectric project on the Congo River, which is primarily used to supply electricity to the Katanfa Copperbelt in the former Belgian Congo, also acted as back-up capacity for the Zambian Copperbelt. This interconnection consequently expanded to provide back-up for both the Zimbabwe and the SAPP grids as a result of the first multiparty power trade agreement.

Electricity was transmitted between Zimbabwe and the Congo, while Zambia received a transmission payment (wheeling charge). The practical experience gained through this arrangement was useful in helping to shape the future establishment of the transmission charges scheme in the SAPP.

Membership of the Southern African Power Pool

There are currently 16 SAPP members, which are listed in Table 2.1. They consist of the following:

- Twelve national power utilities from the continental SADC countries, classified as Operating Members (OP);
- Two Independent Power Producers (IPPs): Hidroeléctrica de Cahora Bassa of Mozambique (which is a Special-Purpose Vehicle; SPV) and Lunsemfwa Hydro Power Company of Zambia; and
- Two independent transmission companies (ITCs), Copperbelt Energy Corporation (CEC) of Zambia and the Mozambique Transmission Company (MOTRACO). CEC is also an IPP.

The majority of these members are national utilities of respective countries operating under the single-buyer model. The aggregated capacity of IPPs is less than 1% of the total. There are nine SAPP OPs, who are also interconnected national utilities. Empresa Nacional de Electricidade of

Table 2.1 SAPP Members in 2017

Full Name of Utility	*Status*	*Abbreviation*	*Country*
Botswana Power Corporation	OP	BPC	Botswana
Copperbelt Energy Cooperation	ITC	CEC	Zambia
Electricidade de Mozambique	OP	EDM	Mozambique
Electricity Supply Corporation of Malawi	NP	ESCOM	Malawi
Eskom	OP	Eskom	South Africa
Hidroeléctrica de Cahora Bassa	IPP	HCB	Mozambique
Lesotho Electricity Corporation	OP	LEC	Lesotho
Lunsemfwa Hydro Power Company	IPP	LHPC	Zambia
Mozambique Transmission Company	ITC	MOTRACO	Mozambique
NamPower	OP	NamPower	Namibia
Rede Nacional de Transporte de Electricidade	NP	RNT	Angola
Societe Nationale d'Electricite	OP	SNEL	DRC
Swaziland Electricity Company	OP	SEC	Swaziland
Tanzania Electricity Company	NP	TANESCO	Tanzania
Zesco Limited	NP	ZESCO	Zambia
Zimbabwe Electricity Supply Authority	OP	ZESA	Zimbabwe

Source: SAPP (2017)

Angola, Electricity Supply Corporation of Malawi, and Tanzania Electric Supply Company Limited of Tanzania are non-OPs which are not yet connected to the SAPP grid. The DRC and Tanzania are also members of the Eastern Africa Power Pool, while the DRC is also a member of the Central African Power Pool.

The SAPP interconnected grid can be broadly divided into three control areas, with Area 1 covering Botswana, Lesotho, southern Mozambique, Namibia, Swaziland, and South Africa; Area 2 consisting of Zimbabwe and northern Mozambique; and Area 3 comprising the DRC and Zambia. Within each control area, one system operator is responsible for balancing the electrical system and managing the power flows between the control areas. These operators are Eskom (Area 1), ZESA (Area 2), and ZESCO (Area 3). The majority of power trades happen within the eastern and central areas.

Electricity markets

The SAPP operates a physical electricity market, which can broadly be divided into long-term and short-term markets, or power exchanges. Electricity is the only product traded on these markets. In the year ending March 2017, approximately 9.0 TWh were traded in all markets, with about 11.3% traded in the power exchange (SAPP 2018).

The market is open to participants who have been licensed or given permission by the host country to undertake cross-border trading, have been accepted as a market participant by the SAPP Executive Committee, are physically connected to one of the SAPP control areas, and have arrangements for provision of balancing services; signed the SAPP Market governance documents; and opened the requisite accounts for trading purposes (SAPP 2017).

Long-term contracts

The majority of electricity is traded via long-term contracts between the member utilities. These contracts are basically over the counter (OTC) contracts which are signed between two parties, with terms agreed outside of the power exchange. The contracts are arranged for a period from one to five years. Although exact contract details are not disclosed, the electricity is priced based on the production cost, but can also vary between consumption periods which can be peak, standard, and off-peak. Naturally, supply delivered at peak hours is priced higher. Trading arrangements are mutually agreed between bilateral parties, and the transmissions paths are secured in advance (SAPP 2016).

The contractual agreements can be firm and non-firm. Firm contracts cannot be interrupted, and hence the buyer is paying a reliability premium. Non-firm contracts can be interrupted with notice. The contractual prices are based on negotiation and thus may not be tied to domestic prices. These contracts usually also include a re-opener clause, whereby parties can revisit the contract terms periodically in view of changes in market circumstances (SAPP 2016).

The forward physical market

The forward physical market (FPM) is a part of the power exchange market operating on the timescale between a week and a month. This trading platform provides shorter-term products for parties involved in bilateral contracts. The FPM provides both monthly (FPM-M) and weekly (FPM-W) products. The monthly product is "primarily baseload generation" delivered at uniform price and output levels for all hours of the month. The weekly market has two products, including on- and off-peak generation. Each of these products must have uniform price and output levels within their respective hours and cannot overlap with the monthly products. Trading in FPM-W is done on a Thursday every week. Forward markets have an auction-trading model similar to that of day-ahead markets discussed next (SAPP 2016).

The day-ahead and intraday markets

The power exchange in the SAPP also has day-ahead and intraday markets, trading short-term products. The markets are based on the price matching auction model. Market participants submit their bid (sales order) or ask (purchase order) onto an exchange platform where the orders are matched automatically and settled at the corresponding matched price and quantity. The intraday market is mainly for trading of imbalances remaining after the day-ahead trading. In 2016, short-term markets had significantly more purchase than supply offers (SAPP 2017).

Market statistics show that the majority of electricity is traded via long-term and forward contracts. This is to be expected, given the short history of the power exchange and that the market consists of very few large players. Since 2013, the share of traded electricity in day-ahead and intraday has been increasing compared to previous years, while the share of long-term contracts has decreased. In 2013, less than 100 GWh of electricity was traded in day-ahead and intraday markets, while in 2014, the volume grew to almost 510 GWh and reached 761 GWh in 2017. Nevertheless, the share of day-ahead and intraday markets in the total market volume is still rather low, constituting only about 6% (SAPP 2017).

The charges for provision of balancing services are calculated within each control zone and are not linked to electricity prices. A new scheme that could link these two components together is currently under study (SAPP 2017).

Transmission allocation and pricing

The transmission capacity allocation in the SAPP is tied to contractual arrangements. Transmission capacity which remains available after settling bilateral contracts is allocated in the short-term market. Transmission capacity is priced in the SAPP with a transit fee which is based on the booked size of transmission capacity and the distance between the supplier and the buyer. The further the transmission distance, the higher the transmission charges. Proceeds from these transmission charges are divided equally among all countries in the assumed wheeling path.

The problem with this scheme is that it prioritizes trading over short distances and does not account for potential transmission bottlenecks (line congestion). The MW-kilometre method also overestimates transmission costs for power "swap" agreements and loads all costs onto the purchaser. Many other markets around the world, such as Europe and Nord Pool, have moved to more advanced methods of transmission capacity pricing such as zonal pricing and implicit transmission capacity allocation. Zonal pricing calculates the price for power transfer based on the available transmission capacity and irrespective of length of the line. In the absence of transmission constraints, transmission prices are based solely on distance. If an area becomes congested, local area or zone prices are used. Zonal prices are determined based on the short-run marginal cost (SRMC) of transmission (i.e. the cost of supplying an additional unit of transmission without increasing the transmission capacity infrastructure). The SRMC of transmission typically includes energy wheeling charges, congestion charges, and cost of energy losses during transmission. Zonal prices can better address problems with transmission bottlenecks compared with the MW-km method. The zone price would increase for congested areas, thereby signalling market scarcity and thus motivates investment into new transmission capacity development in order to alleviate congestion. This methodology also shares the transmission costs more equitably between sellers and buyers. The zonal pricing approach is currently being studied by the SAPP Market Committee (SAPP 2017).

Electricity prices and tariffs

Average market clearing prices are available for forward, day-ahead, and intraday markets. There was a general decrease in market clearing prices in the day-ahead market in 2016/17 (the first such decrease in its history), with an average price of USc 7.23/kWh compared to an average of USc 8.31/kWh recorded in 2015/16. The amount of electricity traded in short- and medium-term markets also decreased over the same period (SAPP 2017).

In general, average electricity tariffs in Southern African countries are very close to, and in some countries even below, the cost of producing and transmitting power. The tariffs in SAPP member countries have been historically low due to government subsidies and, as a result, IPPs may find it difficult to agree on PPA terms with state-owned utility companies that are essentially monopoly buyers. The low-price environment also deters potential IPPs from developing new generation projects without a range of price and non-price guarantees.

There are no available data on transmission prices which are added to the generation cost, although the International Renewable Energy Agency (IRENA 2015) has provided a range of estimates from US$1.4–28/kW per year. The large difference in the estimated range provides an indication of the costs involved with these transmission assets. Despite the extent of this range, the study observed that these transmission price estimates align with similar estimates in international markets. For example, transmission prices range from US$1–23/kW per year in England and Wales and US$2–17/kW per year in Brazil.

Protection of interests

Parties involved in electricity transactions are formally protected by intergovernmental agreements which are discussed next in more detail. These agreements authorize national utilities of member states to participate in the SAPP and also provide security for financial liabilities of buying utilities. If a utility fails to commit to its payment obligations for delivered power, the liability of payment lands on the government of the utility involved. Similarly, the supply utility (sellers of electricity) is allowed to turn to its government should its normal attempts at credit control, such as cutting of supply for non-payment, be circumvented. Although this scheme looks good on paper, the available data suggest that it is not very effective, as many utilities show poor financial performance, high numbers of debtor days, and report inconsistent data.

Current state of supply and demand

Supply mix and capacity

The total installed capacity in the SAPP was approximately 61.9 GW in 2017, while the available capacity was about 54.4 GW and the operational capacity was about 59.5 GW (SAPP 2018). The SAPP annual report estimated the generation capacity surplus at 0.9 GW in 2017. The majority of installed capacity is coal-fired (62%) and located mainly in South Africa and neighbouring countries. Hydropower is the second major source (21%), coming from the Zambezi and Congo basins. South Africa is not only the country with the largest installed generation capacity (accounting for around three-quarters of the total installed capacity in the SAPP) but also the largest electricity exporter accounting for 88% of SAPP net exports in 2016/17. It was also the largest net importer accounting for around 40% of the total.

Electricity demand and consumption

Electricity demand in the SADC region has been increasing in line with ongoing economic growth and has generally been higher than the available generation capacity in recent years. This situation may be even more dramatic in the future as electricity demand continues to grow. However, there is a lack of robust estimates and consensus on the future demand data in the SAPP. Electricity demand forecasts in the SAPP are based on projections prepared by member utilities, most of which are inaccurate and underestimate the actual demand growth. SADC members assumed a 4.4% electricity demand growth rate in the SAPP after considering the electricity access targets agreed under the African Union's Programme for Infrastructure Development in Africa (PIDA). Meanwhile, estimates by the Common Market for Eastern and Southern Africa (COMESA) assume a 7% annual growth rate, which is much higher than the 4.4% assumed by the PIDA. Even though there are conflicting views on the pace of the demand growth, the forecasts are unanimous about the general trend.

While currently the largest share of electricity demand in the SAPP comes from the energy-intensive industrial and mining sectors, looking into the future, increased electricity access in SAPP member countries is expected to become a key driver in the growth of electricity demand. Judging by the poor access to electricity of rural communities, electricity access remains a challenge in the region. IRENA (2015) estimates that the SAPP has the lowest rate of electricity access when compared to other Regional Economic Communities (RECS). The SAPP's electricity access rate stands at 24%, while the East Africa Power Pool and the West Africa Power Pool achieved electricity access rates of 36% and 44%, respectively. While eight of the SADC member countries have electricity access rates of below 30%, in some areas, only 5% of the population has electricity access.

Geographical power flows

Geographical power flows are significant between major supply and demand centres. The mining and manufacturing industries in South Africa,

Botswana, Lesotho, Namibia, and Swaziland and in the Copperbelt are the largest electricity consumers. Over 80% of generation capacity is located in South Africa, followed by hydropower in the Congo and Zambesi basins. Consequently, power flows mostly through interconnectors in the eastern and the central regions. South African electricity exports supplied almost all of Botswana's demand and over half of Namibia's demand in 2016/17, while the current drought has rendered Zambia a net importer over recent years.

Infrastructure bottlenecks

Generation

In its early stages, the SAPP had excess generation and transmission capacity to enable smooth electricity exchange between its members. This situation has changed after years of population and economic growth which were not met with adequate additional generation capacity. Moreover, existing power plants have deteriorated with age and cannot be fully utilized at times due to frequent maintenance requirements. As a result, the region has experienced a supply gap since 2007, which became a local issue for almost every member country. However, industrial recession over the past couple of years has reduced demand and increased levels of self-sufficiency in terms of domestic electricity supply. Table 2.2 shows that most countries now have enough domestic capacity to be self-sufficient, yet in 2015, only Angola and South Africa were in this category. Countries primarily affected

Table 2.2 Available Generation Capacity and Peak Demand of SAPP Members

Country (Utility)	*Installed Capacity*	*Peak Demand*	*Available Capacity*	
	MW	*MW*	*MW*	*% Peak Demand*
Angola (ENE)	2,210	1,599	1,772	111
Botswana (BPC)	892	610	410	67
DRC (SNEL)	2,442	1,359	1,098	81
Lesotho (LEC)	74	140	70	50
Malawi (ESCOM)	366	323	351	109
Mozambique (EDM/HCB)	3,074	1,780	2,279	128
Namibia (NamPower)	508	629	354	56
South Africa (Eskom)	46,963	34,122	4,4134	129
Swaziland (SEC)	61	232	55	24
Tanzania (TANESCO)	1,367	1,051	1,225	117
Zambia (ZESCO)	28,787	2,005	1,385	69
Zimbabwe (ZESA)	2,045	1,521	1,555	102
ALL	**88,789**	**45,371**	**54,688**	**121**

Source: SAPP (2017)

by the supply gap were the hydropower-dependent nations which had no alternative sources in the case of drought. Countries like Zambia and Zimbabwe had to turn to planned load shedding measures for their populations and emergency electricity imports, which come at a high price.

Transmission and distribution

Addressing supply gaps through trade has been difficult due to bottlenecks in the transmission system. The system of competitive trading in spot markets has revealed transmission network constraints, particularly in the eastern and central corridors where the majority of power exchanges take place. The presence of these transmission constraints has negatively impacted cross-border trading, essentially reducing day-ahead and intraday liquidity by close to 40%–50% compared to a scenario absent of these transmission constraints.

Infrastructure development

Infrastructure development in the SAPP is independently undertaken by participating utilities. There were some efforts to coordinate these developments, but progress has stagnated. In 2009, SAPP released a regional master plan with the help of international agencies, estimating the financial benefits of coordinated regional planning. Although this document outlined some significant benefits, promoting coordinated resource planning, regional optimization, and equitable sharing of benefits, it is non-binding and provides flexibility for national governments to decide on the level of autonomy and self-sufficiency. This has made it difficult to establish a coordinated guideline for generation and transmission capacity development.

One of the causes for weak implementation of the master plan has been due to national priorities overriding regional interests. Proponents of national interests argue that major events in the past (like military conflict or drought) have created a need to put national interests of energy security before the regional interests of economy and affordability. Country positions vary between national and regional capacity planning exercises, with more emphasis on self-sufficiency in their national development plans.

Private sector participation in infrastructure development remains very limited in the entire region. The main reasons are non-cost-reflective tariffs, high perceived risks, and weak protection of investor interests. Although this list of issues is noted in numerous publications covering the region, little progress has been achieved so far in attracting private sector investments. A notable exception is the 300 MW coal power plant built by a mining company in Zambia after the country decided to increase electricity tariffs for industry. This was accompanied by a raft of concessions and price assurances that made this an extremely attractive investment for the company concerned.

Organizational structure

The SAPP is a subsidiary institution under the SADC, which evolved from SADCC in 1994. The SADC is committed to regional integration and poverty alleviation in the region. The SADC is made up of 15 member countries, 12 continental nations, and 3 island nations (Mauritius, Seychelles, and Madagascar).

The Directorate of Infrastructure and Services (DIS) is the overarching body through which the SAPP reports to the SADC Secretariat. The SADC DIS heads the organization of the SAPP and is primarily responsible for all energy-related matters, including high-level policymaking relating to the electricity sector. The DIS determines the institutional structure and market conditions for each member country.

The SAPP itself consists of two main committees: the Executive Committee and the Management Committee. Heads of all participating utilities make up the Executive Committee, which also takes on the role of the board of directors of the SAPP. The Management Committee, on the other hand, is made up of senior managers of the participating utilities.

There are four working sub-committees in the SAPP: the Environmental Sub-Committee, the Markets Sub-Committee, the Operating Sub-Committee, and the Planning Sub-Committee. These sub-committees are managed by the Management Committee and are responsible for data collection and the preparation of proposals and reports which the Management Committee would present to the Executive Committee. The Coordination Centre fulfils the secretariat role for the SAPP committees and sub-committees, providing organizational support in terms of keeping minutes and other administrative functions. The centre is also placed in charge of managing energy trading and coordinating multinational projects. These five sub-committees (inclusive of four working sub-committees and a secretariat) are staffed by the respective energy trading departments of participating utilities, while financing for the Coordination Centre is sourced through an annual subscription fee. The interests of the participants are protected by the Intergovernmental Memorandum of Understanding, while enforcement of contract terms is backed by the respective governments (SAPP 2017).

Regulation

The SAPP has an association of regional electricity regulators, the Regional Electricity Regulators Association (RERA), consisting of 10 (out of a total of 12) member state representatives. The RERA is responsible for harmonization of inter-country market rules and procedures that do not fall under national jurisdictions. However, it is important to note that the RERA does not act as the regional regulatory body with regulatory supervision and enforcement roles, but remains a coalition group of national regulators. At present, the RERA has the following strategic objectives:

- **Capacity building and information sharing.** Facilitate electricity regulatory capacity building among members at both national and regional levels through information sharing and skills training.
- **Facilitation of electricity sector policy, legislation, and regulations.** Facilitate harmonized electricity sector policy, legislation, and regulations for cross-border trading, focusing on terms and conditions for access to transmission capacity and cross-border tariffs.
- **Regional regulatory cooperation.** Deliberate and make recommendations on issues that affect the economic efficiency of electricity interconnections and electricity trade between members that fall outside national jurisdiction, and exercise such powers as may be conferred on the RERA through the SADC Energy Protocol.

The RERA's institutional position mirrors that of the SAPP at the time of its establishment. Both parties started on a non-binding cooperative basis from where they were expected to evolve along with the development of the SAPP. At the same time, the institutional roles of the RERA and the SAPP indicate the current shallow stage of the SAPP integration.

Financing of the SAPP

The SAPP is incorporated in Zimbabwe as a non-profit organization. Generation and interconnection infrastructure in the SAPP are owned by the respective national utilities, although some are held under SPVs) established specifically to execute joint projects. One example of a SPV is MOTRACO, which is jointly held by Eskom, SEC, and EDM. Unlike other SPVs, which are typically established by state-owned utilities, the Hidroeléctrica de Cahora Bassa is held directly by the respective governments (Mozambique and Portugal), with the Mozambique government holding majority ownership. The CEC is the only private sector utility and is both an IPP and an ITC.

Investments in cross-border regional infrastructure are typically financed through the national utilities involved, although the SAPP has also been receiving donations from international donors, such as development agencies like the World Bank and the European Investment Bank, among others.

These international donors are also engaged directly through aid to SADC member states. Notably, the Nordic countries were heavily involved in funding cross-border infrastructure projects, such as the Cahora Bassa interconnector and the Zimbabwe and Mozambique transmission networks, that laid the sound foundations for the interconnected South African grid (IRENA 2015).

Daily operations of the SAPP, as well as the working sub-committees, are financed through the SAPP operating budget. All members contribute to the operating budget, although Eskom and ZESA, the largest utility in the SAPP and the host utility of the Coordination Centre, respectively, contribute about 50% of the total budget between them. Recent revisions were made to the contribution shares to ensure a more equitable cost allocation. Specifically, IPPs and ITCs and other members are expected to contribute 5% of the SAPP operating budget. Financing for projects of common interests are also sourced from the member states.

Conclusions

The case study of the SAPP provides an example of relevant experience of cross-border power trading, which started from bilateral agreements between 12 vertically integrated national utilities and later developed as an international spot market without domestic market liberalization. The institutional structure within the SAPP is relatively well-defined and involves all relevant stakeholders. The institutional integrity is grounded by a set of guiding principles and documents that have been built upon mutual consensus to ensure smooth market operations in the pool. SAPP is based on a collaborative approach to market integration with a high emphasis on self-sufficiency and sovereignty of decision making. On the other hand, this approach provides few available instruments to influence developments in member states and to promote harmonization of rules and regulations.

SAPP has a fairly advanced power exchange offering products in monthly, weekly, day-ahead, and intraday markets. The amount of electricity traded in spot markets has been growing, particularly after improvement of day-ahead and intraday markets in recent years. Nevertheless, the share of electricity traded in the power exchange is still low, at about 6% of total electricity traded. This issue could be due to the relatively short history of the power exchange, whereby participants are still inexperienced in short-term trading and require time to adjust. Another cause could be high market concentration, with South Africa's Eskom having over 75% of the total share and trading with 11 other national single-buyer utilities. Such an environment favours long-term contractual commitments. Linked to this point, long-term contracts also allow parties to secure the scarce transmission capacity well in advance. Finally, the low volume of the spot market could be due to its unstable functionality and limited efficiency of matching trade offers.

The most notable challenge for the SAPP is that its generation and transmission facilities are currently unable to provide the SADC region with a reliable electricity supply. Although coordinated infrastructure development is promoted by the SADC community, most of the projects never move beyond the planning phase. Some of the reasons for insufficient capacity development are lack of bankable projects due to electricity subsidies, lack of available project financing, and weak regulatory and institutional arrangements which hurt the investment environment. As a result, despite clear identification and prioritization of a pipeline of projects, commercialization, and operationalization of these projects still experience long delays.

The generation deficit and transmission constraints negatively affect the economies of the SADC countries and contribute to social tensions. Particularly affected are the countries in the northern region which lack alternative sources of supply besides hydropower, which is highly weather-dependant. Infrastructure bottlenecks also limit electricity trading and reduce trust in the benefits of electricity market integration.

References

Beta, M. 2013. *Overview of the SAPP*. Accessible at: www.usea.org/sites/default/files/event-/SAPP%20Overview.pdf.

IRENA (International Renewable Energy Agency). 2015. *Africa Clean Energy Corridor: Analysis of Infrastructure for Renewable Power in Eastern and Southern Africa*. Abu Dhabi: IRENA. Accessible at: www.irena.org/publications/2015/Jan/Africa-Clean-Energy-Corridor-Analysis-of-Infrastructure-for-Renewable-Power-in-Eastern-and-Southern.

SAPP (Southern African Power Pool). 2016. *Annual Report 2016*. Harare, Zimbabwe. Accessible at: www.sapp.co.zw/annual-reports.

SAPP (Southern African Power Pool). 2017. *Annual Report 2017*. Harare, Zimbabwe. Accessible at: www.sapp.co.zw/annual-reports.

SAPP (Southern African Power Pool). 2018. *Official Website*. Accessible at: www.sapp.co.zw.

3 European electricity markets

Introduction

The establishment of a common market for energy has been one of the most significant achievements of the EU. Since the early stages, which date back to the late 1980s, Europe has achieved some important steps in creating the world's largest integrated electricity market, although more work remains to be done. Achieving deeper internal energy market integration is one of five dimensions of EU Energy Strategy, along with boosting energy security, improving energy efficiency, decarbonizing the economy (not least by using more renewable energy), and supporting research, innovation, and competitiveness.[1]

The integrated European electricity market is expected to achieve economic, security, and environmental benefits across Europe. Cost savings will come from avoided generation capacity investments and a more competitive generation landscape. The security benefits will come from an increased choice of electricity suppliers in well-interconnected networks, which are more secure than the isolated markets of individual member countries. Integrating and distributing higher shares of renewable energy will reduce greenhouse gas emissions in electricity generation, which is one of the most GHG-intensive industry sectors in Europe.

This chapter presents an overview of the European electricity market, which is the world's largest initiative of electricity market integration. It covers historical development, market mechanism trends, and the institutional framework that have helped countries to exchange power more freely. Liberalized power sectors and unrestricted grid access have been the cornerstones of the European integration strategy. European experience in the establishment of cooperation bodies between national transmission system operators (European Network of Transmission System Operators for Electricity; ENTSO-E) and regulators (Agency for the Cooperation of Energy Regulators (ACER) and Council of European Energy Regulators

(CEER)) is a particularly valuable model for the ASEAN region. Coordinated infrastructure development (Ten-Year-Network-Development-Plans) and transmission system pricing and allocation are other valuable lessons from the European experience.

Historical development

The vision of a uniform energy market has been one of the main pillars of a common market for goods and services between the EU member states. The first significant steps for market integration were achieved in 1985 and 1986 with the Delors White Paper and the Single European Act, respectively, which set out an ambitious plan to establish a single internal market by 1992. However, neither of these documents had energy on its agenda. The importance of integration of energy markets became more apparent in the late 1980s, but it took another eight years before an agreement on electricity had been reached.

Figure 3.1 highlights the three stages of electricity market integration in Europe between 1996 and 2009. These stages are outlined in more detail in the following sections. It is important to keep in mind that European countries already had well-developed electricity infrastructure by the time the process of market integration commenced. In the early 1990s, most national power sectors operated under state-owned monopolies or single-buyer models, while the first stages of market liberalization were being introduced in the UK.

1996–2003: First Electricity Directive

The first agreement on electricity market integration, the First Electricity Directive (96/92/EC), was signed in 1996 after prolonged discussions

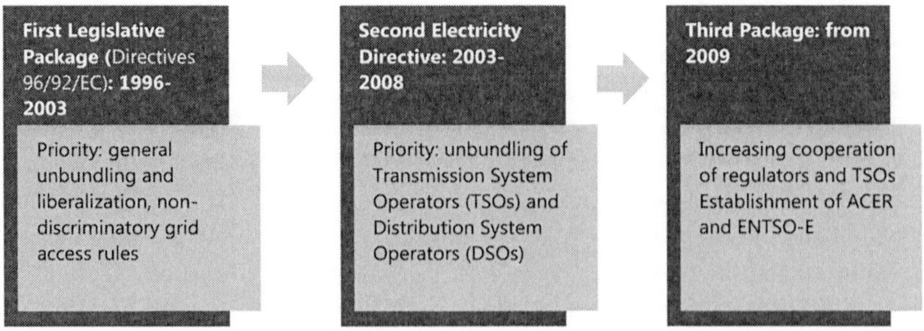

Figure 3.1 Three Liberalization Stages of European Electricity Markets

between EU member states. Member states were required to include the Directive into their national laws by 1998.

The First Directive set out common rules for the internal market in electricity. The aim of the Directive was to guarantee the free movement of capital (i.e. the right to freely establish power companies and construct and operate power stations across the EU) and the free movement of goods to serve final customers. It also sought to unbundle transmission system operators (TSOs) and unbundle commercial operations from generation and distribution. Key features of the Directive were as follows:

- Full liberalization of the generation sector and allowing all large and medium-sized companies to choose their electricity supplier; and
- Access to the grid enabled through unbundling of the vertically integrated state-owned companies and guarantees of non-discriminatory access for all new market players.

However, during negotiations between the European Council, European Parliament, and the European Commission, the complete liberalization of retail markets was watered down and only liberalization of industrial customers was retained. In addition, vertically integrated companies still enjoyed advantages of insider information compared to the independent power producers. Other issues included the absence of cross-border coordination regarding general market design, lack of independence of regulators from the political agenda, and the slow transposition of the Directive by member states into their national law. These shortcomings were addressed in the Second Legislative Package in 2003.

2003–2009: Second Electricity Directive

The Second Electricity Directive came into force in 2003 and was expected to be transposed into the legal system of EU member states by July 2004. The Second Directive was comparable to the First with regard to scope and focused on outlining the design for cross-border markets. According to the Directive, all non-household and industrial customers should be eligible to choose their suppliers, and by July 2007, this right should be extended to all customers.

The overall direction of the Second Directive was for providing more independence to TSOs and DSOs in market operations, as well as to separating regulators from the governments of the member states. With respect to TSOs and DSOs, the Directive demanded at least unbundling the operator in legal terms, which meant establishing a separate legal entity and ensuring that the management was not part of the structures of the integrated company. The Directive also stipulated the establishment of sector-specific regulatory authorities across the EU member states, but since no requirement for political independence was made, the former situation in many member states persisted. The EU regulators were thus not particularly effective in enforcing unbundling of network operators from competitive parts of vertically integrated companies.

Overall, though the first two Directives made progress, weaknesses persisted. The main issues were (1) persistent vertical integration of electricity markets that restricted third-party access to transmission and distribution networks, (2) weak regulatory infrastructure, (3) limited competition as a result of small consumer segments, (4) insufficient liquidity in wholesale markets, and (5) an underutilized interconnection capacity. High access charges, inadequate unbundling of TSOs, and lack of an institution representing TSOs' interests on the EU-wide level also presented challenges for the integration of European electricity markets. Moreover, member states were still found to be protecting their national incumbents.

At the adoption of the Second Directive, the European electricity market consisted of a group of seven regional markets (some of which overlapped) which were physically connected to a certain extent. The countries in Western Europe had electricity exchanges offering trades in both spot and futures contracts. Bilateral (or OTC contracts) represented the majority of trades, although the share of spot trading had been increasing, albeit at different rates in different power exchanges. Some of these markets started to extend their activities to neighbouring countries. Countries in Eastern Europe, however, relied almost entirely on OTC trading with a very low degree of unbundling. In 2017, the total volume of electricity traded on European markets amounted to 12,647 TWh, of which OTC trades accounted for approximately 75% of the total (EC 2018).

From 2009: Third Energy Package

To address the shortcomings of the First and the Second Directives, the EU adopted the Third Package in 2009, which was enforced in 2011. The aim of this package was to improve the functioning of the market and to resolve the key structural problems, such as utility unbundling and the independence of the regulators. Key goals of the package were as follows:

- A complete unbundling of TSOs from generation companies facilitated through either of three pathways depending on the individual preference of EU countries:
    1. Full unbundling requiring generation companies to sell off their gas and electricity networks;
    2. Creation of an independent system operator (ISO), which would be responsible for the operation, maintenance, and investment in

the grid. Provisions still allowed energy supply companies to own parts of the electricity transmission networks, although operational and investment decisions would now be left with the ISO; and

3 Creation of an independent transmission operator (ITO). Provisions maintain the ability of energy supply companies to own and operate gas or electricity networks, but only through a subsidiary, which was required to make operational and capital decisions independently of the parent company.

- Establishment of ACER, an institution representing regulators' interests. This unit would be responsible for publishing non-binding framework guidelines that specify the objectives and principles of the finalized network codes. The codes constitute the integrated market design in Europe.
- Establishment of the ENTSO-E, an institution representing interests of the national TSOs and DSOs. ENTSO-E was established to facilitate cooperation and align interests between TSOs in achieving the efficient use and development of interconnectors. Given significant differences in operating standards between EU states, ENTSO-E was to design common network codes to harmonize the flow of electricity between different transmission systems. It was also to design infrastructure development plans in cooperation with the European Commission and ACER.
- Increased transparency through information disclosure in retail markets to benefit consumers.

Of these, the most significant rules were those on unbundling and establishing of ACER and ENTSO-E. As of 2015, significant progress had been made due to the Third Package, with 96 of the approximately 100 TSOs in Europe being certified as compliant with one of the unbundling models specified earlier. The role of ACER and ENTSO-E was to foster cooperation in cross-border trade and coordinate the new capacity investments, given the large size and complexity of the European electricity markets. It must be stressed that these targets were set without jeopardizing the interests of the national regulators and TSOs who retain the right of the final word within their respective electricity systems.

Coupling of national markets and the "Target Electricity Model"

Over the past three decades, countries in Northern and Western Europe have progressively liberalized their electricity sectors and introduced multiple competitive markets or power exchanges. The "Target Electricity Model" is a framework of the gradual coupling of these exchanges into one entity, operating under a single auction platform. Market coupling is a complex process that requires a series of actions aimed at harmonization of technical, market, and regulatory standards across the market borders. In order to maximize the benefits of market coupling, this process should be accompanied by the development of interconnection capacity that serves as a supporting pillar of the integrated market.

Currently, the most progress in Europe has been achieved in the coupling of the day-ahead markets. The day-ahead markets have been synchronized across Europe to a great extent, using a single price algorithm. This algorithm calculates electricity prices across Europe, taking into account available transmission capacity while still maintaining a certain degree of independent operation within national systems. Day-ahead coupling started between France, Germany, and Benelux in 2010. The approach of determining day-ahead electricity prices in each national market independently has been replaced with a uniform algorithm, calculating prices across the whole region based on the available transmission capacity. As a result, the exchange of electricity went more smoothly, and prices across the regions converged. Another important consequence was that transmission capacity allocation has become an integral part of the price calculation algorithm, while prior to this transmission, prices had been calculated separately.

By 2014, price coupling had been achieved in the day-ahead markets of South-Western Europe and North-Western Europe. These two regions were linked in 2015 and the now-coupled area currently covers 19 countries, representing about 85% of European power consumption (ACER 2016).

Coupling of other markets, such as intraday and balancing, are likely to be the next major milestones for the European market. While coupling of balancing markets remains rather unexplored territory, some progress has been achieved in the coupling of intraday markets under the Cross-Border Intraday (XBID) Market Project initiative, which will be discussed in more detail in the markets section of this chapter.

Geographical coverage

The integrated European electricity market was planned for the entire 28 EU member states. Norway is physically connected and synchronized with the EU member states through the Nord Pool market. Switzerland has not agreed on the terms of joining the European power market, despite the available physical infrastructure. The countries of Albania, Bosnia and Herzegovina, the Former Yugoslav Republic of Macedonia, Kosovo (in line with United Nations Security Council Resolution (UNSCR) 1244 and the International Energy Agency, International Court of Justice Opinion on the Kosovo Declaration of Independence), Moldova, Montenegro, Serbia, and

Ukraine are known as the "contracting parties". This collaboration aims to extend the EU's internal energy market to South-Eastern Europe and the Black Sea region.

Integrating European electricity markets has proved to be a difficult venture. National power sectors in Europe are extremely diverse in terms of their electricity generation mix and had evolved independently before a common vision for market integration had been established. Faced with the slow progress of full market coupling, the EU has established four regional markets that will serve as intermediary stages for the common European market. The four markets are the Nordic region, North-Western/Central Europe, the Iberian Islands, and Southeast Europe.

Vision and objectives

The European electricity market has a multifaceted vision, which at the initial stages was achieving power-sector liberalization and free exchange of electricity between the member states based on principles of market competition. The countries involved are also expecting significant economic benefits arising from market integration, which are discussed later (Newbery, Strbac, and Viehoff 2016).

Faced with global environmental challenges and the new objectives of the EU policy on energy and climate change, the vision of the integrated market has evolved in tandem. According to the EU Commission,

> An internal electricity market for Europe is not an end in itself. It is urgently needed to achieve all of the objectives of the EU's policy on energy. Those include: secure and competitively priced supplies; renewables and climate change targets of 2020 and beyond; and a significant increase in energy efficiency across the whole economy. That market should be based on fair and open competition.[2]

The current vision for the European electricity market is strongly linked with goals of the recently adopted 2030 Climate and Energy Framework and the mechanism of the European Union Emissions Trading Scheme (EU-ETS). The 2030 Climate and Energy Framework sets forth legally binding EU-wide targets and policy objectives for the period between 2020 and 2030. The key aim of this framework is to facilitate the EU in achieving "a more competitive, secure and sustainable energy system as well as to meet its long-term 2050 greenhouse gas reductions target" (EC 2015). The reduction target will be achieved through an emissions trading mechanism, which imposes a hard cap on emission allowances in various sectors. Energy generation plays the pivotal role in emissions trading, as it produces the majority of greenhouse gas emissions in Europe. An important consequence of this package for the European electricity market is that future market design needs to consider its contributions in supporting these policy objectives. Decarbonization and renewable energy targets raise a plethora of other questions, including the optimal balance of baseload power, long-term investments in power generation technologies, idle fossil fuel capacity, variable renewable energy integration, and energy storage. These issues, despite their importance, extend beyond the scope of this book.

Market participants

The European power market consists of the regional markets of EU member states and non-EU members (Norway and Switzerland) which are linked to a different extent. In the north, the European market is linked with Nord Pool, although Nord Pool is commonly treated as a separate market. Market integration has been more advanced in the northern and western regions.

European electricity markets have numerous participants, which can broadly be divided into the following categories:

- National utilities and independent power producers;
- TSOs and distribution companies;
- Power industry regulators;
- System operators;
- Institutions (e.g. ENSTO-E and ACER); and
- Industry groups and market brokers.

Electricity markets

Electricity markets of European countries are widely diverse in terms of operational mode and products offered. The state of vertical unbundling is uneven across Europe, with North-Western Europe reaching the most advanced stage consisting of multiple power exchanges. Trading in South-Eastern Europe is still done predominantly via long-term contracts. The "EU Target Electricity Model" defines four markets: day-ahead, intraday, balancing, and forward markets, that could commonly be developed and linked across European countries in future years.

Trading in electricity markets takes place both in physical and financial markets, as highlighted in Table 3.1. Physical contracts involve the physical delivery of electricity or balancing services, while financial contracts are risk-hedging instruments, which can also be traded by parties outside of the energy sector (e.g. banks and trading houses). Some of the energy exchanges within the EU operate both in the physical and financial markets,

Table 3.1 Products Traded in Physical and Financial Markets

Market Segments	
Physical Markets	*Financial Markets*
Balancing markets Spot markets • Intraday or within-day products • Day-ahead products • Weekend, weekly, and monthly products until the end of the ongoing month Physical forward markets	Derivatives markets • Financial forwards • Futures • Swaps • Options

Source: ACER (2015)

while other exchanges operate in only one of the two market segments. The gradual transition towards the integrated European market is ongoing in the energy-only market, while balancing is undertaken independently by TSOs in accordance with respective national guidelines. To provide better coordination of cross-border balancing efforts, TSOs exchange information in real time, aided by ENTSO-E.

Trading can be done either in OTC markets or in respective power exchanges if that member state is participating in a power exchange. The importance of both channels is discussed in more detail next.

Bilateral contracts

Bilateral, or OTC, contracts have traditionally been the dominant channel for electricity trade at the European level. As stated earlier, OTC deals comprised about 75% of the traded volume in 2017. However, the importance of OTC has been declining over the years, particularly in the north-western region. In addition, the importance of OTC is different among national exchanges today. For instance, while the OTC fraction has been dominant in the UK, it comprised less than a half of total market volume in Nord Pool. OTC trading is either purely bilateral or carried out via a broker. While transactions carried out at energy exchanges are anonymous and screen-traded, brokers' transactions can be handled via public communication channels. Forwards is another example of an OTC instrument which can provide additional flexibility to contract parties.

Power exchanges

Power exchanges, or wholesale markets, trade various products from short-term spot to long-term forward contracts. The market is divided into bidding zones (as opposed to bidding nodes) featuring equal prices within every zone. Any consumer within one zone is allowed to contract electricity with any generator without limitations, such as physical network limits. However, market participants trading electricity with another bidding zone have to consider grid constraints.

Table 3.1 showed physical and financial products traded in wholesale markets. Physical products range from hourly contracts traded in the intraday market to long-term futures. Financial products include various derivatives. A more detailed overview of products traded at European power exchanges can be found in ACER (2014).

Day-ahead markets and coupling of day-ahead markets

Day-ahead markets in Europe operate under the common principle of a uniform clearing price paid to all generators. One of the significant achievements in advancing the integration of European markets is the coupling of the national day-ahead markets that has been ongoing since 2010. As of 2017, day-ahead markets covering 30 out of 42 European borders had been successfully coupled.

Market coupling involves synchronizing the operational mechanics of different electricity markets, with respect to issues such as gate closure times, operational procedures, types of products available for trading, and transmission capacity allocation across borders. The harmonization of such rules and procedures allows for joint market clearing, allowing all bids and requests to be matched automatically through the exchange platform while taking into account available cross-border transmission capacity. This allows the simultaneous inclusion of available transmission capacity in the optimization process, also known as implicit allocation, as opposed to separate or explicit allocation. This process of implicit allocation removes the obligation of market participants to procure cross-border energy and capacity by internalizing the procedure, thereby allowing parties automatic access. Such processes also facilitate the synchronization of energy and transmission markets.

Intraday markets

Due to different definitions of national intraday markets, we summarize all real-time electricity trading happening within the 24 hours as intraday markets. The volume of electricity traded in intraday markets is typically smaller compared to day-ahead and forward markets and mainly consists of covering imbalances remaining after the day-ahead trades. The increasing penetration of intermittent variable renewable energy (VRE) in recent years has caused more uncertainty in short-term output and more trading in some intraday markets.

Coupling of intraday markets is the next major step for the integrated European market. The target is to achieve a degree of market coupling similar to that in day-ahead markets. The integration progress has been slow so far, especially in many countries in Eastern Europe where intraday markets do not exist.

Integrated intraday markets could help optimize the inter-regional power system in the short-term. Currently, expensive resources on the balancing markets account for nearly all of the short-term market adjustments. Integrated intraday markets could change this and yield cost savings, as well as increase the efficiency of the balancing market.

Recognizing the benefits of an integrated intraday market, European power exchanges, with the help from TSOs from 12 member states, launched the XBID Market Project in 2018. The XBID Market Project aims to promote continuous cross-zonal trading, thereby increasing the overall efficiency of intraday trading in the region by creating one integrated European intraday market. Local intraday markets will be linked with the data on available cross-zonal transmission capacity provided by the TSOs, therefore allowing bid and ask orders from market participants from any one country to be matched automatically by a corresponding order in the absence of cross-zonal capacity constraints.

Balancing markets

Balancing is provided whenever there is a mismatch between market supply and demand, and helps to ensure equilibrium, in and near real time. The requirements for balancing services can be very different across Europe due to inherent characteristics of national power systems, available balancing resources, and ongoing electricity trade. Integration of balancing markets is viewed as the final piece of the puzzle for the EU Target Electricity Model of an integrated European electricity market. However, progress in this market has been limited to date, due to challenges as a result of largely different circumstances in existing national balancing markets or, sometimes, the absence of them.

The initial step towards the integration of balancing markets happened in 2012 when ACER first adopted the Framework Guidelines on Electricity Balancing. These guidelines were put in place to provide strong fundamentals for the eventual integration of national balancing markets and the creation of a pan-European electricity balancing market. The target for European market integration envisages cross-border trading of manually activated reserves (replacement reserves). While various market models have been proposed, taking into account the varying degree of harmonization among markets and TSOs, no concrete solution has been agreed upon yet. Potential market structures that were explored include cross-border extension of national balancing mechanisms and bilateral or multilateral TSO-TSO exchanges, among others.

In general terms, balancing services are managed by TSOs, which source their balancing requirements from balancing responsible parties, who can be power generators or demand-response participants. Balancing services are traded in the respective power exchanges or allocated by national TSOs within respective control areas. However, there is little cross-border harmonization of balancing markets, mainly due to disparate market rules governing the balancing market. While ACER expects that most of the operational discrepancies may be resolved through harmonization of market procedures, some inherent challenges with regard to the available balancing resources in member states may be persistent and difficult to resolve.

Geographical power flows

The patterns and magnitude of power flow across the regions provide some good insights into the process of market integration. In Europe, the proportion of cross-border trade volumes compared to national electricity consumption has been increasing since the adoption of the Third Energy Package, indicating progress towards wholesale market integration. Most notable has been the stable exporting status of Central Western Europe (CWE), within which France has a lot of nuclear capacity. A fall in January 2017 was largely due to a combination of nuclear reactor safety tests and low renewables generation, while a fall in November 2017 was largely attributed to high prices within CWE, thus internalizing flows. The direction of some flows and the status of some countries change due to differences in seasonal electricity demand patterns and to growing renewable energy generation in Europe.

The key drivers of the direction of flow have been as follows:

- The level of demand, particularly variable and weather-sensitive demand;
- Levels of plant availability and/or storage capacity;
- Operating costs (SRMC) of conventional thermal plants; and
- Prevailing output levels of priority dispatch renewable capacity.

Cross-border trade volumes

Despite the decline of electricity demand in the EU between 2008 and 2017, there has been a simultaneous increase in cross-border power trade and the market liquidity ratio. The highest traded volume of power in this period took place in CWE markets and in Nord Pool. Nord Pool has been consistently the most liquid market, maintaining market liquidity of above 80% throughout most of this period and even achieving a liquidity ratio of 96% in the fourth quarter of 2013. The volume of electricity traded in the Central and Eastern European (CEE) region has been the smallest in Europe, although it has shown some impressive growth in recent years, with liquidity ratios posting a rise from 6.4% to 21.4%. Consequently, the overall European market liquidity, including both the mandatory and non-mandatory markets, registered growth from 39% to 51% during the same period. This is as a result of both increased power trade volumes and a general reduction in gross inland electricity consumption.

Electricity prices

Along with increased power flows between interconnected regions, the harmonization of wholesale electricity prices is also a good indicator of the degree of electricity market integration.

In general, price convergence has been showing a fluctuating trend, with a recent decline across almost all regions. However, this decline can only be partly attributed to removing bottlenecks in cross-border power flows. The primary cause of price divergence is changes in the electricity generation mix in different countries, largely resulting from nuclear capacity going offline and being replaced by fossil fuels and renewables.

Transmission capacity allocation

Developing a common mechanism for interconnection capacity allocation remains one of the main priorities for the European market. Traditionally, transmission capacity has been allocated through a system of explicit auctions where cross-border interconnection capacity is calculated in the absence of estimates of transmission flow volumes and with little to no consideration for the impacts on neighbouring systems. This approach frequently resulted in TSOs unjustifiably restricting capacity flows due to disparate security standards, thereby negatively impacting the efficiency of interconnections. A significant achievement of market integration is the move towards a system of implicit transmission capacity allocation. In simplified terms, this mechanism instantly includes transmission capacity into the wholesale electricity price calculation within one bidding zone. Implicit transmission capacity allocation has been implemented along with day-ahead market coupling in the European market.

Another refinement of the transmission allocation mechanism is flow-based capacity allocation which minimizes the occurrence of unscheduled flows. These distortive flows occur in complex interconnected systems with many linking points and reduce the available interconnection capacity. Flow-based allocation models provide a more comprehensive solving algorithm which takes into account relationships between multiple interconnectors. It assists, therefore, in managing congestion across borders in the absence of infrastructure upgrades.

To coordinate neighbouring markets close to real time, interconnector transmission capacity must be allocated appropriately to all parties. Under the EU system, transmission capacity is allocated through forward transmission allocation which is conducted years or months before the day-ahead market. Market participants are required to nominate their transmission rights a few hours before market clearance of the spot market. TSOs then use these nominated volumes to estimate the remaining transmission capacity available in the day-ahead and spot markets. The optimal use of all available transfer capacity is thus ensured with market coupling in adjacent markets. Under current circumstances, these transmission rights are not traded in intraday markets, though they may be reserved for balancing in specific circumstances. For longer periods, long-term transmission rights give market participants the ability to hedge against congestion costs.

Implementing a single algorithm for day-ahead market coupling as well as implicit capacity allocation and flow-based allocation of transmissions have improved the utilization of the existing interconnection capacity. The efficient utilization of electricity interconnections has increased from around 60% in 2010 to 86% in 2016 (ACER 2017). "Efficient" use means that whenever there is a price difference between the bidding zones, electricity can freely flow from the zone of low price to the zone of high price, thus converging prices. Although expected in theory, this has not always been the reality prior to implementing of these mechanisms.

The European Network of TSOS (ENTSO-E)

Since the adoption of the Third Energy Package, the role of TSOs has systematically evolved. The liberalization of electricity markets and ongoing market integration required a structure facilitating cooperation and transparency of operations between different TSOs. Some important questions that remained unsolved were harmonization of technical and market standards while increasing the degree of cross-border interconnection.

ENTSO-E was created in 2009 from 6 predecessor organizations and currently contains 43 European TSOs. ENTSO-E is a non-binding institution which operates on a consultative basis. The collaborative nature ensures that national goals of TSOs are not violated. Upon its creation, ENTSO-E was tasked with developing standards and drafting network codes that would later be formally adopted in EU legislative processes to coordinate power flows across different transmission systems and to harmonize investment decisions with regard to new network and transmission capabilities. Network Codes on Capacity Allocation and Congestion Management, Forward Capacity Allocation, and Electricity Balancing provide an exhaustive market procedure handbook that captures market operations for power exchanges (from a year prior to the day-ahead market closure to just before delivery) across the EU electricity market. Such harmonization efforts set the stage for price convergence across markets, as well as for the optimal allocation of transmission capacity through the various national transmission grids.

ENTSOE-E operates through its four sub-committees covering the following functions: (1) legal and regulatory, (2) system development, (3) system operations, and (4) market and research and development. The four sub-committees report to ACER, whose mandate is to ensure that

> market integration and the harmonisation of regulatory frameworks are achieved within the framework of the EU's energy policy objectives. The latter aim to create
>
> - a more **competitive**, integrated market which offers consumers more **choice**;
> - an **efficient** energy infrastructure guaranteeing the free movement of energy across borders and the transportation of new energy sources, thus enhancing **security of supply** for EU businesses and consumers; and
> - a **monitored and transparent** energy market guaranteeing consumers fair, cost-reflective prices and **the deterrence of abusive practices**.[3]

ENTSO-E is also mandated to release a pan-European investment roadmap which highlights key investment gaps every two years. A surprising development has been the creation of multinational network companies, such as Elia and Tennet, which own assets in several member states. The downside is that there is an unresolved conflict of interest, as ENTSO-E is responsible for network planning, while at the same time, their member companies build and operate the assets they have been planning.

ACER and CEER

The EU established the independent ACER to boost cooperation between different national regulators as a part of the Third Energy Package. Rather than functioning as a national regulator, ACER was an attempt to bundle all competencies related to cross-border trade, while other regulatory competencies remained with the national regulators. Through ACER, the EU attempted to (1) reinforce the EU's standing as the key body accountable for all negotiations with third-party countries, (2) legally separate the regulatory functions from the EC and member states to ensure independence and objectivity, (3) simplify the existing regulatory and institutional infrastructure, and (4) bundle technical expertise within EU bodies. The independence of regulators from both industry interests and the government was ensured under the Third Energy Package by creating legal entities at the national level with sufficient funds provided by national governments and authority over their budgets. With the Third Energy Package, regulators had enforcement powers and could issue binding decisions to companies at the member state level and impose penalties in case of non-compliance. Regulators were also empowered with access to data from generators, network operators, and other companies. Cooperation was encouraged between regulators of different EU countries to promote competition; however, this cooperation has been slow to develop (ACER 2016).

Established in 2000, CEER, made up of independent energy regulators, focuses on regulatory issues, while ACER deals more with the legislative issues. CEER also represents Europe in the International Confederation of Energy Regulators (ICER), the umbrella body through which similar regional regulatory bodies (such as the National Association of Regulatory Utility Commissioners (NARUC), America, Energy Regulators Regional Association (ERRA), Central/Eastern Europe, and the Association of Mediterranean Energy Regulators (MEDREG), the Mediterranean region) convene. Established as a "not-for-profit" organization of regulators' collaboration, CEER assists ACER in preparing the yearly market monitoring reports (CEER 2018).

The European Commission

The European Commission (EC) is the executive body of the EU, which represents the collective interests of the EU. Through its Directorate-General for Energy, the EC is responsible for developing and implementing a pan-European energy policy. The Directorate-General works with national governments contributing to achieving three main goals across the EU: security of supply, competitiveness, and sustainability. The EC works closely with ENTSO-E and ACER when implementing technical and market standards for electricity markets and making network codes into EU-wide law.

In developing a European energy policy, the EC ensures that EU-wide targets, such as the 2020 Energy Strategy, are met. The EC also proposes legislation which is then adopted by the co-legislators, the European Parliament, and the Council of Ministers, and enforces the EU-wide law.

Current state of supply and demand

According to Eurostat, total gross electricity generation in the EU-28 in 2016 was 3,255 TWh, which also represents the second year with increased demand after more than a decade of a downward trend. The most obvious reason for falling demand was the economic crisis of 2008–2009. However, as shown next, economic recovery has not caused electricity demand to rebound significantly, suggesting that power demand in Europe could be moving towards saturation (Eurostat 2017).

Although the EU economy recovered (in terms of GDP growth) from the economic crisis, electricity demand in 2016 was still lower than in 2010 and has been rather stagnant over the past six years. While EU electricity consumption decreased between 2010 and 2016 by around 3.5%, GDP increased by almost 6% by the end of 2016 compared to a 2010 baseline. This trend is partly attributed to improved energy efficiency as well as structural changes in the EU economy (Genoese and Egenhofer 2015). Another possible driver is an increase in distributed power generation, which is not always collected in statistics provided by TSOs.

The top four consuming nations in 2016 were Germany, France, Italy, and the UK. A broad declining trend could be observed for the EU-28's electricity consumption, with 18 member states declaring a reduction in electricity consumption from 2010 to 2016. Electricity consumption declined significantly in Portugal, registering a 13.7% decline. The UK and Italy followed closely behind, reporting a 12.2% and 11.4% reduction in electricity consumption, respectively (Eurostat 2017). In contrast, rising net electricity generation was reported for Hungary (63.2%), Lithuania (28.7%), and Poland (23.5%).[4] Other member states with double-digit growth were Ireland and Slovakia.

With regard to the EU-wide electricity supply mix, fossil fuels currently account for a little under half (44.4%) of the EU electricity mix. This is followed by nuclear, which accounted for 25.6% of gross electricity generation in 2017. The share of electricity generated from renewable sources, including large-scale hydro, has reached 30% – a figure that excludes embedded generation. Overall, the share of fossil fuels and nuclear has decreased since 1990, while the share of renewables, including hydro, approximately doubled over the same period (Eurostat 2017).

Governance and policy

The integration of European electricity markets is collaborative in nature. The integration is meant to achieve a greatest common benefit for participating member states while maintaining a certain degree of autonomy of the national power sectors. The institutional influence of the EU in the energy sector has increased significantly since the adoption of the Lisbon Treaty (Article 194) in 2009, which set the overall strategic direction of the EU's energy sector (integration, security, and minimization of environmental impacts). However, even after the adoption of the Treaty, many policy competencies remained at the national member state level. For example, each country is allowed to choose its fuel mix as long as it complies with EU regulations on safety and environmental protection. The progress in EU-wide policy directions requires voluntary cooperation of all participating countries.

Development and implementation of new policy instruments is a complex process. New initiatives must be reviewed by the EC and go through a voting process in the European Parliament before they obtain legally binding power within the EU. ENTSO-E and ACER play a crucial role in advising the EC on infrastructure development and prioritization of interconnections. These institutions are also responsible for the development of technical standards of cross-border market operations, such as network codes and codes of operation. Some of the main policy documents relevant to the electricity market integration include the following.

2030 Energy Strategy

The 2030 Energy Strategy is the main directive shaping the development of the future electricity market that underlines two legally binding initiatives, including carbon dioxide emission reduction and renewable energy and energy efficiency targets. These targets are based on a thorough analysis jointly undertaken by the member states and have been agreed upon. In 2014, the European Parliament voted for the adoption of these targets:

- A binding target to cut emissions in EU territory by at least 40% below 1990 levels by 2030. In addition, member states can have nationwide targets extending beyond these levels. The majority of emission reductions would be achieved through a reformed emissions trading scheme (ETS). The ETS raises important implications, especially for carbon-intensive electricity generation sources, such as coal, diesel, and gas.
- A binding target at EU level to boost the share of renewables to at least 27% of EU energy consumption by 2030.

- On the basis of the Energy Efficiency Directive, the European Council has endorsed an indicative energy savings target of 27% by 2030. This target will be reviewed in 2020, having in mind a 30% target.

Furthermore, the EC has called on member states to achieve a 10% interconnection target by 2020, although this target is not legally binding (EC 2015).

The Electricity Directive

Directive 2009/72/EC relates to the establishment of common rules governing the internal markets in electricity. The Directive not only sets in place harmonized rules for generation, transmission, and distribution activities but also provides clear definitions of key terms with respect to electricity markets. Interestingly, discussion on electricity storage was omitted from this Directive. The Directive also laid down priority dispatch for electricity generated from renewable energy sources (Article 15 (3)) and emphasized transparency in balancing markets. Specifically, Article 37 (6) of this Directive required national regulatory bodies to ensure non-discriminatory and cost-reflective balancing tariffs while also ensuring that relevant incentives are in place to motivate market participants to balance their input and off-takes (EC 2016).

Framework guidelines and network codes

Network codes are grid operation and trading rules that govern the actions of grid operators in terms of transmission capacity allocation. Essentially, these codes govern the decision behind who may use the electricity networks and under what conditions, and are typically commission regulations containing legally binding rules. While such rules were traditionally set independently by the member states, it is evident that a system of EU-wide rules should be drawn up in view of the increasing interconnection between the countries.

Development of network codes is primarily carried out by ACER and ENTSO-E. The overall direction is set by the "annual priority list" prepared by the EC through public consultation and inputs from ACER and ENTSO-E. ACER then develops framework guidelines, which serve as intermediary stage rules. The formal language for network codes is developed by ENTSO-E in alignment with the principles laid down by ACER. A separate Commission decision is required before these network codes become legally binding.

Infrastructure development

Ten-Year-Network-Development-Plans

The Ten-Year Network Development Plans (TYNDPs) are an important set of documents for grid infrastructure development in Europe. The TYNDPs are non-binding in nature – rather, they outline transmission expansion that could bring economic, security, and environmental benefits for the EU.

The process of preparing the TYNDPs is shown in Figure 3.2. In the first step, the analysis scopes four scenarios for the development of the power system based on different fuel mixes and demand projections. Based on these scenarios, 200 experts of 41 TSOs in 34 European countries then carry out bottom-up analysis and identify possible infrastructure bottlenecks under each scenario. A set of additional projects is proposed through agents, which can be either systems operators, who are non-members of ENTSO-E, or electricity storage promoters, or other relevant parties. The results of the planning studies are then combined into a series of infrastructure projects and undergo cost-benefit analysis, as well as environmental and security assessments. ACER is then tasked to ensure consistency between the TYNDPs and existing national plans and policies. ACER may recommend amendments should the need arise. Based on the network development plans, TSOs can decide which projects to undertake. The financing of these expansions comes from national sources.

The TYNDP 2018 projects €114 billion of investments in grid infrastructure, supporting 166 projects in transmission and 15 in storage. It also explores the possibility of a power system where 65% to 75% of the emissions will be cut by 2030 (below 1990 levels) and 80% to 90% by 2040 (ENTSO-E 2018).

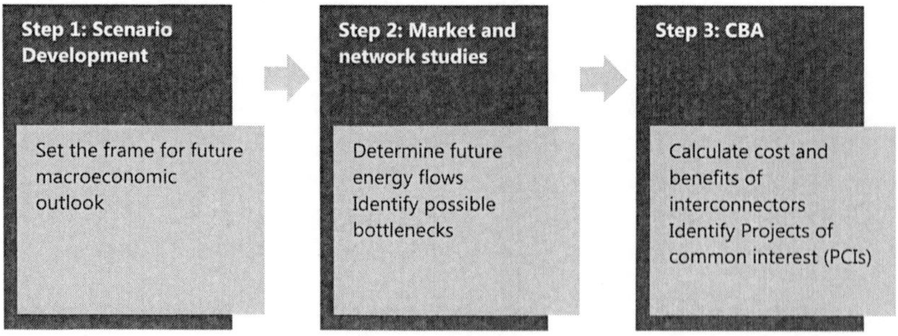

Figure 3.2 Steps of Development of TYNDPs

The importance of TYNDPs is that they align the efforts in infrastructure development between different countries and avoid unnecessary investments that may be carried out independently by member states. However, TYNDPs are not without problems. One particularly difficult area is aligning TYNDPs with a country's energy security goals, especially in light of an unclear trend of natural gas imports from Russia. Moreover, there is a notion that TYNDPs should not be developed by TSOs, who are responsible for building, owning, and operating interconnection projects, and, therefore, could have a vested interest in a particular outcome of TYNDPs. Finally, some projects developed between two or more countries can bring disproportional benefits to the involved parties, or even benefit a third country which does not need to develop any new network. In such a case, the extent to which all affected countries should contribute to the budget of a project cannot be easily agreed upon. Finally, many infrastructure projects face difficulty in receiving social acceptability.

Projects of common interest

Projects of common interest (PCIs) are priority infrastructure developments within TYNDPs which require urgent action from the EC. These projects include the development of new cross-border interconnectors, strengthening existing ones, and helping to integrate renewable energy. Due to the urgency of these developments, PCIs may be afforded both administrative and infrastructural support, as well as public financial support. Specifically, PCIs may benefit from expedited planning and accelerated permit processing, a friendlier regulatory environment, reduced administrative costs as a result of simplified environmental assessment processes, increased public visibility to attract private-sector investment, and access to public funding support totalling €5.35 billion in the period 2014 to 2020. The current PCIs are Priority Corridor Northern Seas Offshore Grid, Priority Corridor North-South Electricity Interconnections in Western Europe, Priority Corridor North-South Electricity Interconnections in Central Eastern and South Europe, and Priority Corridor Baltic Energy Market Interconnection Plan. Details can be found in ENTSO-E (2018).

Economic benefits of market integration in the EU

Several studies quantify economic benefits resulting from ongoing market integration in the EU region. Notable examples are publications by Booz and Company (2013), Newbery, Strbac, and Viehoff (2016), and some of the yearly market monitoring reports by ACER. The studies generally agree that increased market integration and more efficient use of interconnectors will have a net economic benefit. The benefits will stem from price convergence across the regions, fuel cost savings, decreased capital expenditures, and more efficient use of the interconnection capacity.

The analysis carried out by Booz & Company (2013) estimated the benefits of a shallow market integration to be €2.5 billion to €4 billion per year, which is consistent with values in Newbery, Strbac, and Viehoff (2016) and in ACER (2013). About 58%–66% of this benefit has already been achieved due to the level of market coupling as of 2013, especially in the large electricity markets of North-Western Europe and the Nordic region. The remaining 34%–42% is projected to be achieved with the completion of the Target Electricity Model.

Newbery, Strbac, and Viehoff (2016) estimated that by increasing cross-border power exchange volumes from 10% to 15% of EU electricity demand (i.e. about 158 TWh), monetary savings of close to €1.58 billion every year could be achieved, equivalent to 1% of the wholesale demand. Such estimates are under the assumption that the average price difference of €10/MWh persists between the trading points. Considering a multitude of alternative sources of benefits, such as price arbitrage from the coupling of day-ahead and intraday markets and balancing, realizable benefits could be around €2.4 billion per annum.

The theoretical benefits resulting from a deep market integration derived by Booz & Company (2013) could have a much higher magnitude compared to the current state. Deep market integration could achieve benefits ranging from €12.5 billion to €40 billion per year by 2030, or more than ten times their current value. The results show a magnitude of more than 200% which corresponds to a large number of development scenarios tested. Worth mentioning are also high potential benefits of a common market for renewable energy in the order of magnitude between €16 billion and €30 billion a year.

The role of market transparency

The Third Energy Package puts a high value on the transparency of the future market platform. Transparency of information is essential in creating a non-discriminatory environment for all market players, as well as a hedge against market power abuse. With significant efforts from ENTSO-E, transparency has improved significantly. Since 2007, ENTSO-E has been providing access to a range of regularly updated data sets of value to electricity traders and other market participants through its entsoe.net website.

In 2015, as a response to regulation (EU) No 543/2013 on the submission and publication of data in electricity markets, ENTSO-E launched a new central transparency platform: the ENTSO-E Transparency Platform. This new ENTSO-E Transparency Platform replaces entsoe.net (which remains available for consultation) and will publish as much as three times more data than has previously been available to the general public. The ENTSO-E Transparency Platform provides key data sets with regard to six key components of the electricity market, namely: load, generation, transmission, balancing, outages, and congestion management. These data sets can be accessed freely, with continuous updates throughout the day.

Conclusions

The general consensus among stakeholders is that the integration of European electricity markets has delivered many positive results. The 2014 progress report published by the EC noted that cross-border traded electricity between EU countries has increased, wholesale electricity prices have declined by one-third, consumers have increased flexibility to choose their energy suppliers, and various infrastructural bottlenecks are being eased with new interconnectors in the pipelines. A major milestone has been achieved with price coupling of day-ahead markets.

The progress in market integration is reflected by the steadily increasing volumes of cross-border power trades despite the reduction in total demand. Marked improvements in terms of the efficient use of interconnectors have also occurred, recording efficiency improvement from 60% in 2010 to 86% in 2014. Despite evidence of good progress, ACER maintains a cautious optimism towards full market integration, highlighting in their recent assessments the presence of systematic inefficiencies and market barriers. ACER cites sub-optimal allocations of transmission capacity as a result of cross-zonal capacity allocations and/or poor definitions of bidding zones as one of the significant challenges ahead. In addition, ACER recommended continued investment into expanding the interconnectors and transmission network to support trading at congested borders (ACER/CEER 2015). A similar notion was also mentioned in the single-market progress report by the EC (EC 2014). At the same time, some experts argue that infrastructure development through TYNDPs may lead to an unequally developed interconnection capacity. The reason is due to a conflict of interest among the TSOs who are responsible for network planning and simultaneously receive profits from network operations. The social acceptance of large infrastructure projects in the more densely populated European countries also remains an unresolved issue.

The EU initiatives seek to increase the share of renewables in overall electricity supply to 20% and 27% by 2020 and 2030, respectively. Individual member states have deployed various policies, such as subsidies and feed-in-tariffs, to encourage more renewable generation, bypassing the normal market incentives, such as wholesale price signals and the carbon price signal from the EU-ETS. The renewables support policies shift incentives away from electricity generated from conventional sources, which depresses market wholesale prices and reduces utilization rates. Another consequence is higher demand for back-up power from conventional sources as a result of the fluctuating nature of renewables. The reduced operating hours of the conventional power plants thwarts investments in fossil capacity which is needed alongside the renewables to provide baseload and balancing power. The European electricity markets will need to adapt to this development in the coming years by providing some stable long-term price signals for available capacity, even if it does not provide power. Some EU member states have begun experimenting with capacity markets, which allow generators to be compensated for availability at any given time. However, this type of market is currently not foreseen in the EU Target Electricity Model and is subject to national regulation.

Overcapacity in the generation sector, which was caused by the economic crisis, renewable energy development, and improvements in energy efficiency, is another problem for the European electricity market. It is expected that due to falling wholesale prices, more capacity will become unprofitable and will be retired naturally. Linked to EU-ETS price signals, the retired capacity would also likely be the most polluting. However, additional incentives may be required to encourage decommissioning of carbon-intensive power plants if the market fails.

Notes

1 https://ec.europa.eu/energy/en/topics/energy-strategy-and-energy-union
2 Set out in Article 194 of the Treaty on the Functioning of the European Union.
3 https://acer.europa.eu/m/the_agency/Pages/default.aspx
4 Data classification changes accounted for the bulk of these "increases".

References

ACER (Agency for the Cooperation of Energy Regulators). 2013. *Annual Report on Its Activities Under REMIT in 2012*. Brussels. Accessible at: https://publications.europa.eu/en/publication-detail/-/publication/24a12b36-cbd6-11e5-a4b5-01aa75ed71a1/language-en.

ACER (Agency for the Cooperation of Energy Regulators). 2015. *Annual Report on Its Activities Under REMIT in 2014*. Brussels. Accessible at: www.europarl.europa.eu/cmsdata/84735/ACER%20AAR%202014.pdf.

ACER (Agency for the Cooperation of Energy Regulators). 2016. *Annual Report on Its Activities Under REMIT in 2015*. Brussels. Accessible at: www.acer.europa.eu/Official_documents/Acts_of_the_Agency/Publication/ACER%20Annual%20Activity%20Report%20for%20the%20year%202016.pdf.

ACER (Agency for the Cooperation of Energy Regulators). 2017. *Annual Report on Its Activities Under REMIT in 2016*. Brussels. Accessible at: www.acer.europa.eu/Official_documents/Acts_of_the_Agency/Publication/ACER%20Annual%20Activity%20Report%20for%20the%20year%202017.pdf.

ACER/CEER (Agency for Cooperation of Energy Regulators/Council of European Energy Regulators). 2015. *Annual Report on the Results of Monitoring the Internal Electricity and Natural Gas Markets in 2014*. Brussels.

Booz & Company. 2013. *Benefits of an Integrated European Energy Market*. Report prepared for: Directorate-General Energy, European Commission DG Energy. Brussels. Accessible at: https://ec.europa.eu/energy/sites/ener/files/documents/20130902_energy_integration_benefits.pdf.

CEER (Council of European Energy Regulators). 2018. *Official Website*. Accessible at: www.ceer.eu.

EC (European Commission). 2014. *EU Energy Markets in 2014*. Luxembourg: Publications Office of the European Union. Accessible at: https://ec.europa.eu/energy/sites/ener/files/documents/2014_energy_market_en_0.pdf.

EC (European Commission). 2015. *Energy Union Package: Achieving the 10% Electricity Interconnection Target*. Brussels: EC. Accessible at: https://eur-lex.europa.eu/resource.html?uri=cellar:a5bfdc21-bdd7-11e4-bbe1-01aa75ed71a1.0003.01/DOC_1&format=PDF.

EC (European Commission). 2016. Official website. Accessible at: https://ec.europa.eu/energy.

EC (European Commission). 2018. *Official Website*. Accessible at: www.eu.europa.eu/energy.

ENTSO-E (European Network of Transmission System Operators for Electricity). 2018. *Connecting Europe: Electricity*. TYNDP 2018 Executive Report. ENTSO-E. Brussels. Accessible at: https://tyndp.entsoe.eu/Documents/TYNDP%20documents/TYNDP2018/consultation/Main%20Report/TYNDP2018_Executive%20Report.pdf.

Eurostat. 2017. *European Energy Statistics*. Accessible at: http://ec.europa.eu/eurostat.

Genoese, F., and C. Egenhofer. 2015. *Reforming the Market Design of EU Electricity Markets: Addressing the Challenges of a Low-Carbon Power Sector*. Centre for European Policy Studies (CEPS) Task Force Report. Brussels.

Newbery, D., G. Strbac, and I. Viehoff. 2016. "The Benefits of Integrating European Electricity Markets". *Energy Policy* 94: 253–263. doi:10.1016/j.enpol.2016.03.047.

4 Nord Pool

Introduction

Nord Pool is the joint Nordic Power Exchange, which today operates as a part of the European electricity market, but has developed independently from it. Prior to the establishment of a regional power exchange, Norway, Sweden, Finland, and Denmark relied on bilateral trading between national utilities, which also controlled the transmission system. In the early 1990s, the Nordic countries started to implement power sector reforms which were targeted at improving efficiency and the level of competition in the industry. These reforms naturally opened the way for establishing the first multilateral power exchange in the world, the Nord Pool Spot, in 1996.

The Nord Pool Spot market was first opened between Norway and Sweden, and was soon joined by Finland, Denmark, and the Baltic States. Shifting to the market structure has greatly improved the functionality of cross-border power exchange between the Nordic countries. Though there are still issues to be resolved, it should be acknowledged that Nord Pool is the most advanced multilateral power market in the world and serves as an example for other regional actors, most notably the EU.

The Nord Pool model has also attracted significant attention in ASEAN because of its successful transition from a rather simple bilateral contract-based market trading excess power between vertically integrated national utilities to a vibrant power exchange and a financial market with hundreds of participating actors.

This chapter presents an overview of the Nord Pool market, including its historical background, market structure, policies, and regulation. Given the ongoing integration of the Nord Pool with European power exchanges, and the fact that all Nord Pool members except Norway are members of the EU, some sections in this chapter are rather brief, and the reader is referred to the previous chapter in this book in order to avoid duplication.

Historical development

This section presents an overview of Nord Pool's historical development, which is broadly divided into two phases, before and after the opening of the power exchange.

The old structure

Nord Pool can trace its roots as far back as the 1970s, with the creation of the Nordic Council of Ministers. The oil crisis highlighted to the Council the importance of energy policy, forcing each country to adopt and develop a robust national energy framework and promoting the sharing of available energy resources between Nordic countries.

At that time, despite differences in structure, ownership, and regulation, the Norwegian, Swedish, and Finnish electricity markets all had an oligopolistic structure, with a strong presence of state-owned utility companies that also controlled the national grids. Denmark had two national grids for geographical reasons, each with a similar organizational structure.

Trading of electricity between the countries was conducted through long-term bilateral contracts. The coordinating body was Nordel, an organization set up to promote regional cooperation between the large utility companies. The motivation behind cross-border power trade was the achievement of optimal investment and dispatch of an interconnected system. Therefore, investment decisions on interconnection capacity were based on cost saving potential arising from pooling available generating capacity. As such, member countries frequently exchanged cost information and power trades were settled at the average of the two marginal costs. The energy ministers considered the region a whole market and commissioned studies to standardize and remove existing technical barriers to boost trade volumes.

In 1990, Norway passed a new Energy Act, which paved the way for deregulation of the domestic power industry in order to make it more efficient. Third-party access was afforded to all transmission networks, while integrated utility companies were required to legally separate their generation, distribution, and retail activities. As a result, the main Norwegian utility company unbundled, leading to the establishment of two new companies: Statkraft (for generation) and Statnett (for the main transmission grid and central system operation). Unbundling was the key building block of the reform programme to address the oversupply of electricity in Norway and the resulting market inefficiencies. Sweden, Finland, and Denmark also started the unbundling process in the 1990s, facing dissatisfaction from the private-sector companies and consumers over restricted grid access.

An interesting note here is that unbundling of the national companies in Norway, Sweden, Finland, and Denmark happened without the privatization process. Instead, the new market-forced incumbents to adapt their operations to a new competitive environment which caused large internal restructuring and improvement in operations. Some commentators have suggested that retaining and reforming the incumbents, as opposed to breaking them up, has helped the Nordic utilities to become among the most competitive in Europe.

Creation of Nord Pool

In 1996, a joint Norwegian-Swedish power exchange was established under the name Nord Pool ASA. Nordel, which was a coordinating body between the TSOs in Denmark, Finland, Iceland, Norway, and Sweden, assessed the requirements needed to develop the Nordic electricity market and examined neighbouring developments in the EU and Baltic regions. In 1998, Nord Pool became the first multinational common electricity market, with inclusion of Finland, followed by West Denmark in 1999, and East Denmark and Estonia in 2000.

Nord Pool's post-2000 agenda has been on smoothing the market operations and implementing common standards for cross boundary electricity markets. The percentage of power traded in day-ahead and intraday markets has been steadily increasing over this period. Given the ongoing development of European electricity markets, Nord Pool participants sought to develop stronger ties with the latter and jointly worked on the issues of congestion management and transmission capacity allocation. At the same time, given the more advanced stage of Nord Pool compared to the rest of the EU, the Third Energy Package did not interfere with Nord Pool's regulations.

ENTSO-E officially replaced Nordel, the organization representing the interests of Nordic TSOs, in 2009, when the organization was wound up and all of its operational responsibilities were transferred to the former. In 2013, Nord Pool was joined by Latvia and Lithuania, and in 2014, Nord Pool Spot took sole ownership of the UK electricity market. In the same year, the North-Western Europe day-ahead price coupling project was pioneered by Nord Pool, along with three other power exchanges and 13 TSOs.

Vision and objectives

Nord Pool is the most advanced electricity market in Europe and has a vision of "Pioneering European Power Markets". Nord Pool's primary objective is to ensure the efficiency, transparency, and security of the electricity market, and to provide reliable and low-carbon electricity at competitive prices to customers in seven countries (Nord Pool Spot 2016).

Geographical coverage

It is common to differentiate the region into the Nordic states (Norway, Sweden, Finland, and Denmark) and Baltic States (Lithuania, Latvia, and Estonia). Nord Pool also operates the spot market in the UK and has operations in ten European countries, inclusive of the aforementioned countries. Nord Pool has offices in all countries in which it operates. Each office has its own functions, but the main office remains in Norway, where 50% of the workforce is based. Finland is the second-largest office, hosting most of Nord Pool Spot's IT department and organizations.

Nord Pool also assists and supports other power exchanges that were subsequently launched. Some examples are the German and French energy markets: Nord Pool provided the software and capacity building through Nord Pool Consulting.

Market principles

One of the key successes of Nord Pool is that it operates based on principles rather than on clear cut rules which gives market participants more flexibility. Detailed rules can then be defined and maintained in market participant agreements, which include much more information. As shown next, the focus of the principles is on transparent and open market operations and protection of key information which gives a participant a competitive advantage. Some of the key points include (Nord Pool Spot, 2016):

- Neutrality and non-discriminatory behaviour of the market participants, including ensuring impartiality of all parties and efficient access to information that is of importance to price formation;
- Interest protection of the participants through collateral and settlement obligations that ensure confidence and predictability for the parties;
- "Reasonable" profit from the organization and operation of the marketplace for the licensee;
- Licensee's duty to disclose information to National Regulatory Agencies (NRA);
- Primary capital requirements;
- Market surveillance; and
- Confidentiality of information concerning market participants' business which would be of competitive importance.

Market participants

The Nordic power markets can be divided into two distinct markets: the physical electricity market and the corresponding financial market, which makes it very similar to the European power exchange.

Electricity markets

Nord Pool operates the electricity markets and officially sold off control of the financial market (formerly known as Eltermin) in 2010 to NASDAQ OMX Commodities. The electricity market has both a day-ahead (Elspot) and an intraday (Elbas) market. Participants in the electricity market are as follows:

- **Grid owners.** Grid owners build, operate, and maintain the grid in their specified area. They are responsible for submitting hourly transmission capacities to the TSOs to set the area prices.
- **TSOs.** TSOs are mainly responsible for balancing the grid by ensuring smooth market operations in the day-ahead market and balancing market and handling unexpected contingencies in the real-time market. The TSOs for the Nordic, Baltic, and UK grid are Statnett SF for Norway, Svenska Kraftnät for Sweden, Fingrid for Finland, Energinet.dk for Denmark, Elering for Estonia, Litgrid for Lithuanian, AST for Latvia, and National Grid for the UK.
- **Regulators.** Nord Pool has NordREG, which is the umbrella organization of all Nordic energy regulators, and is the overall body responsible for the Nordic electricity market. Besides regulating the institutional framework and ensuring the efficiency of the electricity market, NordREG is also seemingly responsible for ensuring the efficiency of the financial market for electricity. To illustrate this, NordREG released a report in 2010 assessing the degree of efficiency in the Nordic financial electricity market and considered measures for improving the functioning of the market. The report considered issues of liquidity, competition, and transparency. A more detailed description on the role of NordREG will be provided later in this chapter.
- **Other market participants.** Other market participants include generators, retailers, traders, and larger end-users.

Elspot has an estimated 360 members (buyers and sellers) (Nord Pool 2016). Generators include public utilities and other power generation companies who operate both in the wholesale market and the electricity markets. Retailers serve smaller end-users who do not have access to the wholesale electricity markets, while larger end-users may decide to purchase their electricity directly in the electricity market. Traders are entities that take positions in the electricity markets to gain profits on market mispricing, but who otherwise have no stake in the markets.

Financial markets

NASDAQ OMX Oslo ASA is the official financial exchange platform for electricity derivative products in Nord Pool. The market participants in the financial market can be divided into the following categories (Nord Pool 2016):

- **The clearing house.** One of the key participants in a standardized financial market would be the clearing house. The official clearing house for the Nord Pool financial derivatives market is NASDAQ OMX Stockholm AB, a subsidiary under NASDAQ OMX Oslo ASA.
- **Financial regulators.** NASDAQ and its subsidiaries are responsible for both the exchange trades and over the counter trades. Under the Exchange Regulation ('Børsforskriften') governing NASDAQ OMX, it is required to have an internal regulating and monitoring body that supervises trading activity, among other duties, which is the Market Surveillance department. The Market Surveillance department's role is to maintain market transparency by reporting all OTC trades and disclosing all inside information, and to monitor, highlight and investigate behaviour in violation of Market Conduct Rules, such as insider trading and market manipulation. Furthermore, NASDAQ OMX is also placed under the financial regulation of the Financial Supervisory Authority of Norway, Finanstilsynet, while the clearing house is supervised under the Swedish Financial Supervisory Authority, Finansinspektionen. These independent bodies ensure that both the trading and clearing are in alignment with trading and clearing rules set by the Norwegian Ministry of Finance and the Swedish Financial Supervisory Authority, respectively. Compared to its counterpart NordREG in the electricity markets, the financial regulators mentioned in this paragraph are responsible solely for the trading of the financial instruments and do not ensure liquidity and efficiency of the market, which is NordREG's mandate.
- **Other market participants.** Other market participants in the financial markets can be divided into players from the electricity sector who wish to hedge their risk exposure (e.g. generators, retailers, and large end-users) and the financial market actors who wish to make a profit based on speculative activity (e.g. banks and trading companies). While some

market players are engaged directly in the financial market through membership agreements with Nord Pool, a larger proportion participate through a representative or a broker. In 2017, there were about 360 market players in the NASDAQ market for electricity derivatives in Nord Pool.

The Nord Pool financial market remains a largely regional market, with Nordic participants currently dominating the financial market, hosting about 86% of the market players.

The electricity market

The electricity market in Nord Pool offers a broad variety of products. Since the establishment of Nord Pool Spot as the first multinational exchange for trading electricity, the majority of trading of electricity in Nordic countries has now shifted away from bilateral contracts to power exchange trading.

Bilateral contracts

The importance of bilateral or OTC contracts has been decreasing over recent years. Available market statistics show that the participants are gradually shifting away from OTC to spot market trading. This trend is common to many places with established power exchanges and increasing confidence in spot markets. As of 2017, the Nordic electricity market has the lowest percentage share of OTC in Europe. This is very different from the early 2000s, when the majority of the electricity generated was traded through bilateral OTC contracts. Currently, the price point of the Nord Pool Spot is valued and used as an important reference point in determining OTC contract prices.

Day-ahead market

The Elspot has a day-ahead spot market which is synchronized with other European electricity markets under the day-ahead price coupling scheme. Elspot has 24 one-hour trade contracts in a single day. The Elspot is a double-bidding, day-ahead market, where buyers and sellers submit hourly bids the day before actual delivery, although bidding can start up to 12 days ahead. The system price is calculated at the equilibrium point (where demand meets supply) for each of the 24 hours. The system price is announced close to an hour after bidding closes and contracted volumes are to be physically delivered at the designated intervals accordingly the next day. To accommodate for transmission bottlenecks, hourly zonal market prices for specific areas may be artificially raised to reduce demand and ease transmission constraints. Zonal prices are set by using the system price (price if there were no congestions), coupled with power transmission capacities submitted by transmission service operators on the day before trade. This is known as the implicit capacity allocation and is described in the European markets chapter in more detail. It should be noted that Nord Pool was one of the pioneers of this mechanism.

The total amount of electricity traded in Nord Pool day-ahead markets (including the UK) was 512 TWh in 2017 (Nord Pool 2018). For comparison, the total electricity generated in Europe in 2017 was 3,597 TWh (ENTSO-E 2018).

Intraday market

Elbas is the intraday market covering the Nordic, Baltic, UK, and German markets, and operates in 24-hour resolution. The intraday market supplements the day-ahead market through evening out short-term imbalances between supply and demand in the power market. Capacities available for trading in the intraday market for the next day are announced at 14:00 CET daily. Trading occurs continuously and closes one hour prior to delivery. Price are settled on a best price principle, whereby the highest buy price and lowest sell price are settled first. With increased penetration of wind energy, intraday markets are expected to trade more and more electricity in the future.

Balancing market

Nord Pool launched a synchronized balancing market between Norway, Sweden, Finland, and Denmark in 2008, following final agreement on cross-border TSO cooperation and subsequent approval by the Nordic governments. This agreement was preceded by years of feasibility studies carried out by Nordic regulators. Eventually, the TSOs agreed on the following harmonization measures (NordREG 2010):

- Common gate closure for offers to the regulation power market and for final production and trade plans;
- Harmonized cost base for balance settlement; and
- Common model for the settlement of imbalances.

Denmark, Finland, and Sweden first introduced these measures in early 2009, followed shortly after by Norway, which put them in pace in late 2009. The main differences in balancing markets between the common

Table 4.1 Difference in Balancing Markets between Nordic and European Markets

	European	*Nordic*
Balancing	per country	for the entire Nordic region
Based on	difference between scheduled and measures tie-line interchange	frequency difference
Controller	one per country/TSO	one for the entire Nordic region
Objective	return difference with scheduled within 15 minutes	keep the frequency within the 49.9–50.1 Hz band
Activation time	5–15 minutes	120 seconds (hydro)

market of Norway, Sweden, Finland, and Denmark, and that of Europe are summarized in Table 4.1.

The financial market

Prior to 2010, Nord Pool operated the financial exchange (Eltermin) that hosted the majority of electricity derivatives. In 2010, Nord Pool transferred the ownership of Eltermin to NASDAQ OMX, which now retains the ownership of the financial market for Nord Pool derivatives. However, volume turnover in the Nordic market has been decreasing since 2011. Factors such as high hydropower reserve levels, low wholesale prices, and moderate expectations on economic growth are cited as reasons for the reduced interest in electricity market hedging activities over recent years. However, despite such downturns in market activity, the volume traded on the financial market was still 2.7 times the turnover of the physical market in 2017 (but down from 4.7 four years earlier).

OTC market

Power derivatives can be traded through brokers in OTC markets. Clearing is done at Nord Pool Clearing for exchange-listed products and at NASDAQ OMX for non-listed contracts.

Electricity futures

Products listed on the exchange include Nordic, German, Dutch, and UK power derivatives, such as base and peak load futures, Deferred Settlement Futures, and options. These exchange-listed contracts with settlements ranging from daily, to weekly, monthly, quarterly, and annual contracts, cover positions for up to ten years. The system price from Nord Pool Spot is used as the reference price for settlement of electricity future. Being a financial market, there is no actual delivery of electricity, and as such, transmission costs are not taken into account during settlement. Instead, cash settlements based on the positions taken are conducted throughout the trading and/or delivery period, based on the characteristics of the financial contract.

Transmission-related futures

For end-users and power generators, additional transmission-related congestion costs are hedged using the Electricity Price Area Differentials (EPAD), formerly known as Contract for Difference (CfD) contracts. The EPAD provides a hedge for the difference between system price and the electricity price from a specific area due to constraints in transmission.

Geographical power flows and trade statistics

The geographical power flows between the Nordic countries have daily and seasonal patterns, which are heavily dependent on the availability of the renewable energy sources in the region. In the rainy seasons, Norway and northern Sweden have excess cheap hydropower, which is exported to neighbouring countries. During drier years, southern Sweden and Finland provide stable thermal baseload power, exporting it to Norway and Sweden. Denmark obtains a lot of wind power from its large offshore capacity which is shared with its neighbours.

Nord Pool has multiple interconnectors with other European countries with the total capacity of over 6,000 MW, and the Nordic market generally has been a net exporter of electricity to other parts of Europe over the past three years since 2014.

Electricity prices

A comparison of wholesale electricity prices in Nord Pool with those of other European markets shows that Nord Pool has been consistently the lowest price benchmark in Europe, which is intuitive considering the large amount of hydropower and wind energy generation in the region. Overall, significant price declines have also due to falling fossil fuel prices and generation overcapacity in Europe, in addition to the high influence of weather on wholesale price trends in Nord Pool. The information on zonal price development was provided in the European markets section of this book.

Transmission pricing and allocation

Nord Pool consists of 16 interconnected pricing areas or bidding zones. Transmission pricing in the Nordic market is done through zonal market splitting and implicit capacity allocation, which is now also common in many parts of Europe. The market splitting mechanism allows transmission costs to be priced implicitly into the electricity pricing. The congestion charge is determined as the difference between the area price and the system price.

The Nordic market is split into various bidding areas that may become separate price areas in the presence of transmission constraints in the spot market. In the absence of such capacity constraints, the spot system price converges with the spot price throughout the entire Nordic Power Exchange Area.

One potential shortcoming of this mechanism is that since the transmission service operator retains the congestion earnings, there may be a lack of incentives for investment in grid infrastructure, the issue that has been already mentioned in the TYNDPs section of Chapter 3.

Ownership structure

Nord Pool is owned by the national TSOs who also have the ownership of the physical transmission infrastructure in the member states. This is intuitive because the market relies on information input and scheduling by the TSOs. Specifically, the actual ownership list of Nord Pool is the Nordic TSOs Statnett SF (28.2%), Svenska Kraftnät (28.2%), Fingrid Oy (18.8%), Energinet.dk (18.8%), and the Baltic TSOS Elering (2%), Litgrid (2%), and Augstsprieguma tikls (2%).

Distribution of ownership is a reflection of how Nord Pool has evolved. Norway and Sweden were the founding members with equal ownership majority. Finland and Denmark joined later, at about the same time, and received smaller shares because the market had grown since its establishment. Finally, the Baltic States joined much more recently.

Early success in the derivatives market significantly boosted the value of Nord Pool by the time Finland and Denmark became owners. To lower the cost of ownership, Nord Pool was split into the physical market (owned by all TSOs) and the financial market (originally owned by Statnett and Svenska Kraftnät). However, the split created challenges in managing two different entities: Nord Pool Spot (physical market) and Nord Pool ASA (financial market). Nord Pool ASA was sold to NASDAQ in 2008.

NordREG

Nord Pool has set up Nordic Energy Regulator (NordREG), a regional cooperative body between the Nordic energy regulatory authorities, which oversees the institutional and legal frameworks within the Nordic electricity market. At the European level, NordREG works in close cooperation with ACER and CEER, while most Nordic regulators, except for Norway, share memberships in all three entities.

The establishment of NordREG was a result of a traditionally close cooperation on energy between the Nordic countries. The regulatory role of NordREG covers the wholesale, retail, and transmission aspects of the electricity market. Specifically, NordREG works towards creating a liberalized retail market, an efficient wholesale market, and ensuring reliable transmission and distribution of electricity throughout the electricity market as their priority objectives. To that end, NordREG holds public consultations, commissions research reports, and releases various publications on pertinent issues to achieve their strategic objectives. Beyond ensuring efficiency of the electricity market, NordREG is also responsible for maintaining the efficiency of the financial market.

Current state of supply and demand

The Nordic states have a total installed capacity of 102 GW. Approximately 50 GW of this capacity is hydropower located mainly in Norway and Sweden. Other important sources by capacity are wind and nuclear energy, with both having about 12 GW.

Over 50% of the total electricity supplied in the Nordic states comes from hydropower, given ample water resources in any year. However, annual hydropower generation in Norway and Sweden can vary by 20% based on weather conditions. Another 20% of electricity supply is covered by nuclear power, while 14% and 7% are contributed by combined heat and power (CHP) and fossil fuel generators, respectively. With increasing wind generation and the low marginal cost of production of wind, around 7% of electricity demand is supplied from that source. The Baltic States are much more reliant on fossil fuel generation, which contributes 66% of the total electricity supply there.

Seasonal and temperature changes are the main determinants of electricity demand patterns in the Nordic countries, with colder winters registering correspondingly higher electricity consumption.

Total electricity demand in the Nordic group of countries was approximately 412 TWh in 2017, of which industry, households, and services consumed around 41%, 28%, and 21%, respectively. The remaining 10% was accounted for by agricultural use (2%), traffic use (1%), and network losses (7%). Similar to the rest of Europe, electricity demand in the Nordic states recently stagnated after a period of decline caused by the economic crisis of 2008 and a subsequent rebound.

Market transparency

The success of the Nord Pool scheme is largely due to its high transparency, reflecting that equal information access for all participants improves trust and efficiency. Nord Pool provides a large database of market information that can be accessed on its website. It is also a part of the Regulation on Energy Market Integrity and Transparency (REMIT) initiative by the EU. This initiative requires market participants to submit timely records of their wholesale energy market transactions to ACER through a selected Registered Reporting Mechanism. Norway is also a voluntary participant in this scheme, even though Norway is not a member of the EU.

Other examples which must be reported to REMIT include changes in planned power outages or any other information that can affect market prices significantly. Information must be disclosed to the market or the player must be suspended from trading during the relevant time period. The Nord Pool Spot Market Surveillance Team follows up on them too.

Functional linkage with European markets

Nord Pool is becoming increasingly connected with other European electricity markets. Integration projects are currently ongoing, with the help of the EC, ACER, CEER, and system operators (ENTSO-E). The EU's Energy Strategy has been widely endorsed and adopted by the Nordic countries.

The most significant achievement in market integration between Nord Pool and the European markets is the price coupling of day-ahead markets. Nordic member states are also part of the XBID initiative to achieve intraday market coupling across Europe.

Furthermore, the Nordic TSOs actively participate in the development of TYNDPs and PCIs as a part of ENTSO-E functions. The Nordic regulators have also been active in the creation of framework guidelines and the drafting of opinions of ACER on proposed network codes. Finally, Nord Pool provides consulting services to other electricity markets within and outside of Europe.

Conclusions

The Nord Pool market between Norway, Sweden, Finland, and Denmark is arguably the most successful international power market in the world. The market has well-developed physical and financial electricity markets, and large trading volumes in both market segments. The 2016 market progress reports by ACER highlights that Nord Pool Spot exchange had the second highest amount of traded electricity in Europe in 2015, second to only the European Electricity Exchange Spot. This is despite significantly lower electricity generation and consumption of Nordic countries compared to Western Europe. Even though Nord Pool is exposed to weather-related uncertainties in renewable energy output, as well as the uncertainty with natural gas imports from Russia (particularly in Finland), electricity prices between the Nordic and the Baltic States have been converging, indicating the benefits of their interconnection. With the operationalization of the NordBalt Lithuania-Sweden electricity connection, electricity prices in Latvia and Lithuania experienced a significant drop, nearly converging to the Estonian price level. The interconnection has also resulted in falling carbon dioxide emissions in the Baltic region.

Nord Pool's trading system is being constantly upgraded with new features focusing on efficiency and easy terms of use in order to deliver an efficient trading platform for all Nord Pool users, regardless of their size or geographic location.

Nord Pool has pioneered the concept of cross-border electricity market coupling and coupling of balancing markets in Europe. This expertise puts Nord Pool in a position to implement and test systems for establishing day-ahead markets in other countries, with Bulgaria and Croatia being the most notable examples.

An interesting experience gained from Nord Pool is that the scheme had multilateral cross-border trading before establishing the Nord Pool Spot market. Thus, reforming the power sector by separating the generation and transmission sectors was not a pre-requisite for a multilateral market, which is also evident from the SAPP experience. Nevertheless, unbundling played a crucial role in the later success of Nord Pool, particularly in terms of competition and efficiency. As the market became increasingly mature, more players realized the benefits of having an integrated electricity market, spurring higher trading volumes. In order to promote liquidity in the spot market, all available transmission capacity between different areas was given to the market. Furthermore, regardless of their size, all market participants were required to source a minimum amount of electricity from the exchange.

Current challenges for Nord Pool include the ongoing integration with European electricity markets, particularly with Western Europe, and the transfer of its operational standards to the Baltic States. Other issues include common Nordic (and European) grid planning, provision of balancing services for renewable energy integration, and alignment of interests between international TSOs. The greatest challenge regarding transmission system planning concerns the uncertainty of future technical developments, particularly with renewable energy and energy storage.

References

ENTSO-E. 2018. *Annual Report 2017*. Accessible at: www.entsoe.eu/publications/general-publications/annual-report/.

Nord Pool Spot. 2016. *Official Website*. Accessible at: www.nordpoolspot.com.

Nord Pool Spot. 2018. *Annual Report 2017*. Accessible at: www.nordpoolgroup.com/message-center-container/Annual-report/.

NordREG. 2010. *Harmonising the Balancing Market: Issues to Be Considered*. Report 5/2010. Sweden: NordREG. Accessible at: www.nordicenergyregulators.org/wp-content/uploads/2013/02/NordREGreport5_2010_Balancing.pdf.

5 Power interconnection in the ASEAN region

The case for energy cooperation in ASEAN

ASEAN has ten member countries: Brunei Darussalam, Cambodia, Indonesia, Lao PDR, Malaysia, Myanmar, the Philippines, Singapore, Thailand, and Vietnam. These countries are widely diverse in terms of geography, economy, and culture. Table 5.1 gives some key statistics of these sovereign states. If ASEAN were considered as a single economy, the combined GDP of US$2.6 trillion in 2016 would place it as the seventh-largest economy globally. In absolute terms, Indonesia, Thailand, and Malaysia have the highest GDPs in the region, while GDP per capita is highest in Singapore, Brunei, and Malaysia. The general notion about the ASEAN region is that it is expected to rapidly grow over the coming decades and potentially become the world's fourth-largest economy by 2050.

The combined population of ASEAN member states is estimated to be about 8.7% of the global population. ASEAN consumed 4.5% of the total primary energy in the world, while producing 5.7% of total global energy in 2016. Total electricity consumption in ASEAN has increased from roughly 180 TWh in 1990 to over 837 TWh in 2016 (IEA 2017). Two factors behind this trend have been the ongoing population and economic growth, as well as the increase in energy consumption per capita. As the region's economies continue to grow, its per capita consumption and electricity access are still below worldwide average, creating the likelihood of an even higher pace of growth in demand, both for the individual countries and the region as a whole. Forecasts by the IEA, shown in Table 5.2, indicate that the region's electricity consumption could increase between two and three times between 2016 and 2040, depending on the policy scenario chosen.

Fortunately, ASEAN member states have abundant primary energy resources to address these needs, although they are distributed very unevenly. In the north, Lao PDR, Myanmar, Cambodia, and Vietnam possess large hydropower resources, potentially more than they require domestically; while Indonesia, Malaysia, and Brunei have significant reserves of fossil fuels.

Table 5.1 Key Statistics in ASEAN (2016)

Country	*GDP (Million/ US$)*	*Population (thousands)*	*GDP/Capita (US$/Capita)*	*Urbanization Rate (%)*	*Area (Thousand km²)*
Brunei Darussalam	11,206	423	26,493	78	5.8
Cambodia	19,194	15,156	1,266	21	181.0
Indonesia	931,216	258,705	3,600	54	1,913.6
Lao PDR	15,903	6,621	2,402	40	236.8
Malaysia	299,632	31,634	9,464	75	331.4
Myanmar	68,636	52,917	1,297	35	676.6
Philippines	311,453	103,243	3,017	44	300.0
Singapore	296,977	5,607	52,963	100	0.7
Thailand	407,408	67,455	6,034	52	513.1
Vietnam	196,196	92,695	2,138	34	331.2
ASEAN	2,600,000	634,500	4,034	49	4,500.0

Note: GDP refers to Gross Domestic Product at current market prices.

Source: ASEAN (2017)

Table 5.2 Projected Electricity Demand in ASEAN

		New Policies Scenario		*Current Policies Scenario*		*Sustainable Development Scenario*	
	2016	*2025*	*2040*	*2025*	*2040*	*2025*	*2040*
Primary energy demand (Mtoe)	643	806	1,062	823	1,133	732	891
Share of fossil fuels (%)	74%	76%	76%	77%	79%	76%	59%
Final consumption (Mtoe)	453	560	714	570	752	500	584
Electricity demand (Twh)	837	1,214	1,997	1,247	2,105	1,181	1,828
Energy intensity of GDP (2016 = 100)	100	81	59	82	63	73	49
Carbon intensity of power (2016 = 100)	100	94	83	98	96	79	26

Source: © OECD/IEA 2017 World Energy Outlook Special Report: Southeast Asia Energy Outlook, IEA Publishing, Licence: www.iea.org/t&c, as modified by Anthony David Owen

Virtually all countries have significant wind and solar potential. However, utilizing these reserves may be a challenging task, because not all of these reserves are located near consumption centres or can be easily extracted. Gas-fired generation can be developed quickly, is relatively clean and efficient, but is considered relatively expensive. Coal-fired power generation is affordable, but carries risks of air pollution and greenhouse gas emissions. Hydropower has a big cost advantage over other sources, however it requires large capital investments at the initial stage and exposes its users to seasonal droughts. It can also cause severe political and social tension around water management and use issues. Furthermore, countries may wish to have a diversified portfolio of fuel sources for security reasons. All these arguments present a compelling case for ASEAN to coordinate utilization of the region's available resources and to integrate energy systems between member states.

History of regional cooperation in energy

Energy cooperation in the form of an electricity interconnection in the region began in 1966, when an agreement was signed between Thailand and Lao PDR. This was one year before ASEAN was established as an entity by the five founding members: Indonesia, Malaysia, the Philippines, Singapore, and Thailand. The interconnection project between Thailand and Lao PDR was built in order to supply Thailand's growing energy needs and to boost cooperation between the two countries. The next major milestone in regional energy cooperation took place 15 years later, after the military conflicts in the Indochinese Peninsula had ceased. In 1981, the Heads of ASEAN Power Utilities Authorities (HAPUA) was established to work on a blueprint for electricity interconnection in ASEAN. The ASEAN Energy Cooperation Agreement, signed in 1986, outlined the main areas of potential energy cooperation and proposed an ASEAN-wide electricity network. However, very few concrete steps were achieved in progressing the regional interconnection until the mid-1990s (AEMI 2016).

Since 1995, the key initiatives in energy cooperation of the ASEAN member states have been formalized in the ASEAN Plan of Action for Energy Cooperation (APAEC), which is released every five years.

APG and TAGP

In 1997, ASEAN leaders agreed on the "ASEAN Vision 2020" which aligned interests of the member countries under one common vision for ASEAN. One of the areas emphasized in the document was the establishing of energy networks, spanning across all ASEAN member states and consisting of two main components: the ASEAN Power Grid (APG) and the Trans-ASEAN Gas Pipeline (TAGP). Since publishing the ASEAN Vision 2020, both components have become the de facto main strategic areas of regional energy cooperation and have been included in all subsequent APAECs and AEC Blueprints. Although the ASEAN Vision 2020 identifies main strategic goals for market integration in the region, it does not specify a particular market model for the regional electricity market.

The APG is an initiative to link the member states in a single network in order to enhance electricity trade, provide a means to address the growing electricity demand in the region, to enhance the use of clean energy, and to enhance access to electricity across the region.

The Memorandum of Understanding (MoU) on the APG was signed by the ASEAN member states in 2007. This MoU recommended that member states cooperate in assessing common policy, regulatory and technical standards for cross-border interconnection, and trade. The cooperation was to be promoted on either a bilateral or a multilateral basis and had to comply with national laws of each member country. The areas of work included review and assessment of the required institutional frameworks, at both national and regional levels, to facilitate regional electricity trade; the commercial and economic feasibility of interconnectors; and the construction, financing, operation, and maintenance of the APG. However, this MoU did not contain any concrete milestones to be achieved or information about the targeted market integration model.

Parties responsible for implementation of the APG are Working Group Two (APG/Transmission) of HAPUA and the ASEAN Power Grid Consultative Committee (APGCC). The harmonization of regulatory practices and technical standards is undertaken by the ASEAN Energy Regulators' Network (AERN), established in 2012. The implementation of the MoU is reported by the HAPUA Council through periodic reports, submitted to the senior energy officials in ASEAN.

Implementation and current status

The development of physical interconnection capacity follows recommendations outlined in the ASEAN Interconnection Master Plan Studies (AIMS). These documents help to design an optimal interconnected system by achieving the lowest total system costs, subject to technical, policy, and resource constraints of the member states. The detailed methodology can be found in any of the AIMS. AIMS are prepared by HAPUA and regional experts. Before the first AIMS was conducted in 2003, HAPUA had planned to build APG from 14 bilateral and multilateral interconnection projects which included existing interconnections of Thailand-Lao PDR and Peninsular Malaysia-Singapore. However, the AIMS I, completed in 2003, concluded that it was uneconomic to create a single ASEAN grid and recommend 11 bilateral interconnectors to be constructed by 2019 instead.

Following the re-organization of HAPUA in 2004, its Working Group Four embarked on a second study (AIMS II), which was published in 2010. Unlike its predecessor, AIMS II concluded that an ASEAN-wide power grid was economically viable. However, it recommended the creation of three geographically separate interconnected sub-systems before integrating them into one APG. In addition to the 5 interconnections that already existed at that time, the report listed another 12 projects that were classified as "committed" and another 17 as "generic" (AIMS Working Group 2010).

Following recommendations outlined in AIMS II, HAPUA developed an integration strategy that was aimed at developing bilateral and multilateral interconnections between single member states initially, progressively extending to a sub-regional reach and eventually developing into a fully integrated Southeast Asian power grid. Both AIMS I and AIMS II suggested three sub-systems as intermediate steps to the APG (Roesli 2006):

- System A (Upper West System), consisting of ASEAN countries under the GMS, including Cambodia, Lao PDR, Myanmar, Thailand, and Vietnam;
- System B (Lower West System), consisting of countries under the Indonesia, Malaysia, and Thailand and Indonesia-Malaysia-Singapore Growth Triangle Sub-region, and Indonesia (Sumatra, Batam) and Malaysia (Peninsular and Singapore); and
- System C (East System), consisting of countries under the Brunei Darussalam-Indonesia-Malaysia-Philippines East ASEAN Growth Area Sub-region, including Brunei, Malaysia (Sabah, Sarawak), Indonesia (West Kalimantan), and the Philippines.

The geographical coverage of these systems and list of priority projects are shown in Figure 5.1.

Interconnection capacity development

By the end of 2016, 11 interconnections between 6 pairs of countries were in commercial operation, with a total capacity of nearly 3,500 MW, as shown in Table 5.3.

Table 5.4 summarizes another 13 projects identified in the AIMS II report which are expected to be completed by 2020. Large capacities will be constructed in the Mekong region. There is also a proposed link between Sumatra and Peninsular Malaysia. Most of these projects are currently two years or more behind the original schedule and may not be completed by 2020 (AEMI 2016).

The vision beyond 2020 has over 20,000 MW of interconnectors, of which the links between Thailand and Myanmar take more than 60% (Table 5.5).

Trade volumes

Over the past 25 years traded electricity in ASEAN has experienced a gradually increasing trend. The majority of the traded electricity has been exchanged between GMS members. Lao PDR has grown into a major net

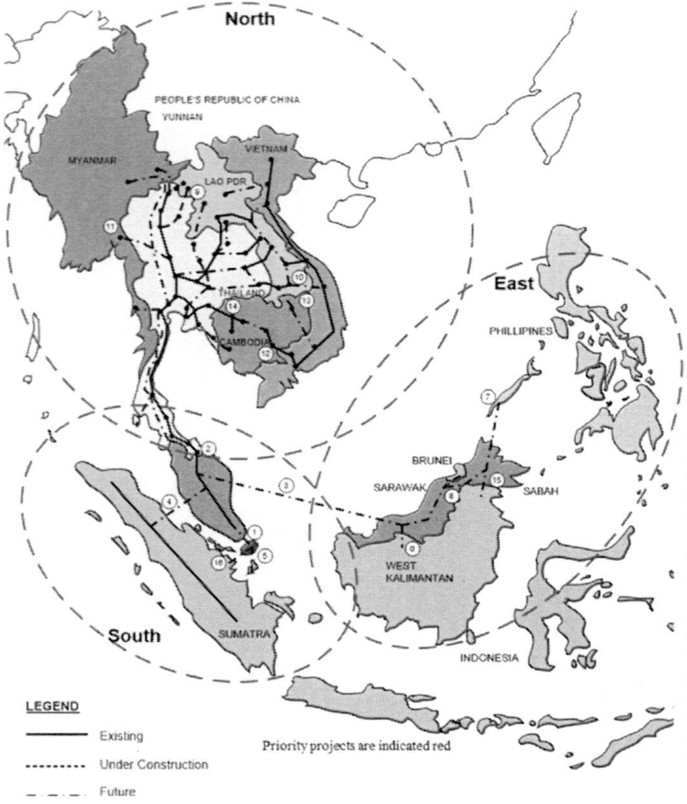

		Earliest COD
1)	P.Malaysia - Singapore (New)	post 2020
•	Plentong - Woodlands	Existing
•	P. Malaysia – Singapore (2nd link Plentong - Woodlands)	post 2020
2)	Thailand - P.Malaysia	
•	Sadao - Bukit Keteri	Existing
•	Khlong Ngae - Gurun	Existing
•	Su Ngai Kolok - Rantau Panjang	Future
•	Khlong Ngae – Gurun (2nd Phase, 300MW)	Future
3)	Sarawak - P. Malaysia	2025
4)	P.Malaysia - Sumatra	2020
5)	Batam - Singapore	2020
6)	Sarawak - West Kalimantan	Existing
7)	Philippines - Sabah	2020
8)	Sarawak - Sabah – Brunei	
•	Sarawak –Sabah	2020
•	Sarawak – Brunei	2019
9)	Thailand - Lao PDR	
•	Roi Et 2 - Nam Theun 2	Existing
•	Sakon Nakhon 2 – Thakhek – Then Hinboun (Exp.)	Existing
•	Mae Moh 3 - Nan - Hong Sa	Existing
•	Udon Thani 3- Nabong (converted to 500KV)	2019
•	Ubon Ratchathani 3 – Pakse – Xe Pian Xe Namnoy	2019
•	Khon Kaen 4 – Loei 2 – Xayaburi	2019
•	Nakhon Phanom – Thakhek	2015
•	Thailand – Lao PDR (New)	2019-2023
10)	Lao PDR - Vietnam	2016-2020
•	Xekaman 3 - Tranhmy	Existing
•	Xekaman 1 – Pleiku 2	2016
11)	Thailand - Myanmar	2018-2026
12)	Vietnam - Cambodia (New)	
•	Chau Doc – Takeo – Phnom Penh	Existing
•	Tay ninh – Stung Treng	Future
13)	Lao PDR - Cambodia	Post 2018
14)	Thailand - Cambodia (New)	post 2020
•	Aranyaprathet – Banteay Meanchey	Existing
15)	East Sabah - East Kalimantan	post 2020
16)	Singapore – Sumatra	post 2020

Figure 5.1 ASEAN Interconnection Projects

Source: HAPUA, personal communication

Table 5.3 Cross-border Interconnectors in ASEAN as of 2016

Project	*System*	*Type*	*Original COD*	*Current COD*	*MW*
P. Malaysia–Singapore					
Plentong–Woodlands	HVAC: 230 kV	EE	–	1958	450
Thailand–P. Malaysia					
Sadao–Chuping	HVAC: 132/115 kV	EE	–	1980	80
Khlong Ngae–Gurun	HVDC: 300kV	EE	–	2002	300
Thailand–Lao PDR					
Nakhon Phanom–Thakhek–Theun Hinboun	HVAC: 230 kV	PP: La->Th	–	1998	220
Ubon Ratchathani 2–Houay Ho	HVAC: 230 kV	PP: La->Th	–	1999	126
Roi Et 2–Nam Theun 2	HVAC: 230 kV	PP: La->Th	–	2010	948
Udon Thani 3–Na Bong–Nam Ngum 2	HVAC: 500 kV	PP: La->Th	–	2011	597
Nakhon Phanom 2–Thakhek–Theun Hinboun (Expansion)	HVAC: 230 kV	PP: La->Th	2012	2012	220
Lao PDR–Vietnam					
Xekaman 3–Thanhmy	HVAC: 230 kV	PP: La->Th	–	2013	248
Vietnam–Cambodia					
Chau Doc–Takeo–Phnom Penh	HVAC: 230 kV	PP: La->Th	–	2009	200
Thailand–Cambodia					
Aranyaprathet–Banteay Meanchey	HVAC: 115 kV	PP: La->Th	–	2007	100
			Total:		**3,489**

Notes: P. Malaysia refers to Peninsular Malaysia. COD refers to Commercial Operations Date. HVAC refers to high-voltage alternating current. HVDC refers to high-voltage direct current. EE and PP refer to economic exchange and power purchase, respectively. La and Th refer to Lao PDR and Thailand, respectively.

Source: AEMI (2016)

Table 5.4 Planned Interconnector Projects by 2020

Project	*System*	*Type*	*Original COD*	*Current COD*	*MW*
Thailand–P. Malaysia					
Su-ngai Kolok–Rantau Panjang	HVAC: 132/115 kV	EE	2014	TBC	100
P. Malaysia–Sumatra					
Melaka–Pekan Baru	HVDC: TBA kV	PP: Sm-> Pb & EE	2015	2020	600
Sarawak–W. Kalimantan			–		
Sarawak–Sabah–Brunei	HVAC: 275 kV	EE	2012	2015	230
Sarawak–Brunei	HVAC: 275 kV	EE	2012–2016	2018	2 × 100
Thailand–Lao PDR			–		
Mae Moh 3–Nan 2–Hong Sa	HVAC: 500 kV	PP: La->Th	2015	2015	1,473
Udon Thani 3–Na Bong–Nam Ngeip 1	HVAC: 500 kV	PP: La->Th	2017	2019	269
Ubon Ratchathani 3–Pakse–Xe Namnoi	HVAC: 500 kV	PP: La->Th	2018	2019	390
Khon Kaen 4–Loei 2–Xayaburi	HVAC: 500 kV	PP: La->Th	2019	2019	1,220
Lao PDR–Vietnam					
Xekaman 1–Ban Hat San–Pleiku	HVAC: 500 kV	PP: La->Vn	2011–2016	2016	1,000
Nam Mo–Ban Ve	HVAC: 230 kV	PP: La->Vn	2011–2016		TBC
Luang Prabang–Nho Quan	HVAC: 500 kV	PP: La->Vn	2011–2016	2020	1,410
Lao PDR–Cambodia					
Ban Hat–Stung Treng	HVAC: 230 kV	PP: La->Kh	2011	2017	300
			Total:		**7,192**

Notes: P. Malaysia refers to Peninsular Malaysia. W. Kalimantan refers to West Kalimantan. COD refers to Commercial Operations Date. HVAC refers to high-voltage alternating current. HVDC refers to high-voltage direct current. EE and PP refer to economic exchange and power purchase, respectively. Pb refers to Pekan Baru. Sm refers to Sumatra. La refers to Lao PDR. Th refers to Thailand. Vn refers to Vietnam. Kh refers to Cambodia. TBC is an abbreviation of "to be confirmed". TBA is an abbreviation of "to be advised".

Source: AEMI (2016)

Table 5.5 Planned Interconnection Projects beyond 2020

Project	*Type*	*Original COD*	*Current COD*	*MW*
P. Malaysia–Singapore	PP: PM->Sg	2018	Post–2020	600
Thailand–P. Malaysia	EE	2016	TBC	300
Sarawak–P. Malaysia	PP: Sw->PM	2015–2021	2025	4 × 800
Batam–Singapore	PP: Bt->Sg	2015–2017	2020	3 × 200
Phillippines–Sabah	EE	2020	2020	500
Sarawak–Sabah–Brunei	PP: Sw->Sb	2020	2020	100
Thailand–Lao PDR	PP: La->Th & EE	2015–2023	2019–2023	1,000+
Lao PDR–Vietnam	PP: La->Vn	2011–2016	TBC	TBC
Thailand–Myanmar	PP: Mm->Th	2016–2025	2016–2026	13,000+
Vietnam–Cambodia	PP	2016	TBC	TBC
Thailand–Cambodia	PP: Kh->Th	2015–2017	Post–2020	2,200
E. Sabah–E. Kalimantan	EE		Post–2020	TBC
Singapore–Sumatra	PP: Sm->Sg	2020	Post–2020	600
		Total:		**22,274–25,424**

Notes: P. Malaysia and PM refer to Peninsular Malaysia. E. Kalimantan refers to East Kalimantan. E. Sabah refers to East Sabah. COD refers to Commercial Operations Date. HVAC refers to high-voltage alternating current. EE and PP refer to economic exchange and power purchase, respectively. Sg refers to Singapore. Sw refers to Sarawak. Bt refers to Batam. Sb refers to Sabah. La refers to Lao PDR. Vn refers to Vietnam. Mm refers to Myanmar. Th refers to Thailand. Kh refers to Cambodia. TBC is an abbreviation of "to be confirmed". TBA is an abbreviation of "to be advised".

Source: AEMI (2016)

exporter of hydropower over this period, quadrupling its electricity exports from 2000 (2.8TWh) to 2013 (12.5TWh). Thailand currently accounts for four-fifths of Laotian electricity exports (80%), with Lao's neighbouring countries receiving the remaining 20%. Thailand's economic growth has been the main contributor to near quadrupling of its electricity imports between 2000 and 2013. This increase in electricity trade between the two countries has been largely facilitated by the operation of the Nam Theun 2 Hydropower station. Formerly a net electricity exporter in the 2000s, Malaysia currently imports electricity not only to meet rising demand, but also to maintain the stability of its power grid in times of emergency.

As a region, Southeast Asia is viewed as an electricity importing region, as a result of rapid economic growth over the past decade. The majority of electricity imports into Southeast Asia have been supplied by China.

Implications

At its current stage, the APG is mainly focused on the development of the physical interconnection capacity between single member states. There has been little progress in other areas, such as development of complimentary institutional, legal, and commercial frameworks for an integrated electricity market. Some regional experts argue that this is because there is no clear vision for the fully developed APG. A deeply integrated APG could, for example, include a competitive electricity market open to all participants (such as the EU market). Such a model would require ASEAN countries to develop comprehensive institutional, regulatory, and market arrangements, which in turn require a very high level of cooperation and trust between the member states. A shallow version of APG could be a physical interconnection scheme with a limited or even absent market framework which links several heterogeneous grids and markets in the region (such as the power pools in Africa).

While it is recognized that progress with the APG has largely been stalled, progress can be seen from the successful implementation of the following steps (Shi and Malik 2013):

- Adoption of the AIMS I, and the updated AIMS II, thereby laying down guiding principles for future implementation of the ASEAN interconnection projects;
- Signing of an MoU on the APG, thereby laying down foundations for collaborative efforts towards APG;
- Restructuring of HAPUA and the formal establishment of a permanent, but rotating, HAPUA Secretariat, to facilitate streamlining of operations; and
- Establishment of the APGCC as an umbrella body to supervise the development and implementation of APG projects.

As a result, by 2015, the ASEAN region had the following links established between the electricity systems of member countries:

- Cambodia imports electricity from Thailand, Vietnam, and Lao PDR;
- Indonesia (West Kalimantan) imports electricity from Malaysia (Sarawak);
- Lao PDR imports from and exports to Thailand and Vietnam;
- Malaysia imports from and exports to Thailand;
- Singapore is interconnected with Peninsula Malaysia, but only for purposes of energy security;
- Thailand imports from and exports to Lao PDR and Malaysia, while only exporting electricity to Cambodia; and
- Vietnam imports electricity from Lao PDR and China, and exports electricity to Cambodia and Lao PDR.

Benefits to participating countries

General benefits

The benefits of creating an ASEAN-wide interconnection system range from cost and fuel savings to enhanced energy security of the countries and environmental benefits for the region. The increased geographical coverage and integration of a common grid will allow all member states to take advantage of a variety of available resources. A very often cited example in the context of ASEAN is development of hydropower resources in the Mekong region and sharing them with neighbouring countries in the south. Displacing the use of hydrocarbons with cheaper hydropower could benefit these growing economies with reduced electricity costs and lower greenhouse gas emissions.

At least in theory, the potential benefits derived from existing and planned electricity interconnectors could be substantial. Intuitively, countries with large hydropower potential – namely Myanmar, Lao PDR, and Cambodia – are expected to sell electricity to key demand centres such as Thailand, Malaysia, Vietnam, and Singapore. The inclusion of Singapore, however, is problematic, not only due to its current 50% overcapacity in its generation sector but also its strategic desire to remain self-sufficient. Due to its strategic location, Thailand may become an electricity trading hub, through which electricity from Myanmar, Lao PDR, and Cambodia could be exported to Malaysia. Furthermore, mutual benefits could be reaped if interconnectors could be established between Malaysia and Indonesia, wherein Peninsula Malaysia could import electricity from Indonesia and Sarawak could export electricity to Indonesia.

Quantified benefits

The analysis done by AIMS II compared systems with and without interconnections, which were identified as a part of the AIMS list. The results are shown in Table 5.6. New interconnections could save a total of

Table 5.6 Cost Savings of Interconnection Projects Identified in AIMS II

	Total Expenses (Million US$)		*Savings*
	Without Interconnection	*With Interconnection*	
Costs structure	83,699	81,980	1,719
Capacity costs	253,025	252,871	154
Fuel costs	336,724	334,851	1,873

Source: AIMS Working Group (2010)

US$1.87 billion, of which US$1.72 billion would be capacity costs and US$154 million would be fuel costs.

Challenges for interconnection projects

A range of obstacles have delayed the integration progress of electricity sectors in ASEAN. ASEAN countries vary greatly in their size, landscape, level of economic development, and national energy resources. Countries like Indonesia and the Philippines are archipelagos, separated by large bodies of water, while mountain ranges in Indochina, Sumatra, and Borneo create immense technical and economic challenges for infrastructure development. ASEAN member countries also vary considerably in terms of power sector regulations, market structure, and technical characteristics. All of these circumstances create barriers to effective regional energy cooperation.

Coordination of infrastructure development

In theory, large interconnected systems need less reserve capacity for the same level of system reliability compared to disconnected power systems. In the case of power shortages, an interconnected power grid allows each region to also draw supply from other regions instead of relying solely on domestic generation capacity. This ability reduces required investments in the power sector infrastructure and creates economic benefits in the form of cost savings. However, in order to reap these benefits, countries would need to give up a certain degree of self-sufficiency in domestic supply and rely on power exchanges instead.

While this is a logical step for market integration that has been achieved in many other parts of the world, developments remain nascent in ASEAN. Currently, most national power development plans in ASEAN countries give priority to domestic power generation when planning for the future. The trend of interconnectors being built on a localized per project basis is expected to prevail, as there appears to be little to no enthusiasm for developing a coordinated approach to building cross-border transmission lines. A relevant example would be the GMS trading scheme described in the case study that follows. Even though power trading is growing in the GMS region, the contracts remain purely bilateral and often involve developing infrastructure retained solely for these contracts. In addition, power generation projects are often barred from selling power to entities other than the contracted party, either through physical transmission lines or through power purchase agreements.

Different technical standards

Different technical standards of power system operations between ASEAN member states is another limiting factor. Although countries have common technical standards domestically, these standards can vary greatly across countries. Currently, there is still no set of common technical standards for ASEAN. Existing interconnection projects have not addressed these issues appropriately. For example, the power exports in GMS are mainly exported through dedicated transmission infrastructure for specific projects, independent of national grids. Furthermore, technical standards can be relatively easily aligned for these cases because power flows are small and unidirectional. However, with developing regional power systems, there will be a necessity for regional coordination of technical standards. Technical failures would have a much more pronounced impact in large interconnected systems compared to small-scale lines.

Access to finance

Access to finance remains another important challenge. According to the IEA, the ASEAN region would need to spend US$618 billion in generation and US$690 billion in the transmission and distribution sectors to meet its rising demand to 2040. Compared to that, the cost of APG in its current state is estimated to be "only" US$20 billion (IEA 2015).

It is still unclear how the financing for all these projects is going to be procured. While some interconnections are expected to be commercialized due to financial backing from multilateral development banks, bilateral agencies, and the private sector, many others appear to lack financial viability, despite the societal benefits they may bring. Given that the provision of interconnectors may be perceived as a public good, there remain significant barriers in attracting private sector interests to finance such projects.

Institutional differences

ASEAN member states have diverse electricity markets, displaying a spectrum of power sector characteristics, institutional, and administrative features that may hinder interconnection due to large technical and operational discrepancies. While Brunei, Cambodia, Lao PDR, and Myanmar still adopt a vertically integrated market structure with traditional state-owned utilities, Indonesia, Malaysia, Thailand, and Vietnam have introduced upstream competition by allowing private IPPs in the power generation sector, while downstream transmission remains largely nationalized. Currently, only the

Philippines and Singapore have unbundled their electricity market, with separation of the generation and transmission sectors. Due to the diversity between these regional electricity markets, even if funding is available, electricity grid interconnections are complex to develop and manage. Table 5.7 clearly illustrates the diversity seen between ASEAN member states in terms of power industry structure.

The tendency of ASEAN member states to implement domestic reforms, rather than consult with other member states, has been a major barrier in achieving closer cooperation. One example could be national regulations on interconnections that make it difficult to justify an interconnection with another country. Moreover, individual countries tend to place emphasis on self-sufficiency rather than cooperation due to domestic energy security concerns. The disparity in terms of tariff structure, electricity subsidies, as well as taxation and other institutional procedures such as the sequencing of approval procedures, further complicate the investment climate. These trade and investment barriers create a harsh investment environment, particularly for private investors, who may find them difficult to navigate.

Table 5.7 Power Sector Structure in ASEAN

Country	*Regulator*	*Regulator Independence*	*Market Structure*
Brunei Darussalam	Dept. of Electrical Service	Under the Ministry of Energy	Single buyer
Cambodia	Electricity Authority of Cambodia	Independent	Single buyer
Indonesia	Dept. of Energy and Mineral Resources	Under the Ministry of Energy and Mineral Resources	Single buyer
Lao PDR	Dept. of Electricity	Under the Ministry of Energy and Mines	Single buyer
Malaysia	Energy Commission	Independent	Single buyer
Myanmar	Ministry of Electric Power	Under the Ministry of Electric Power	Single buyer
Philippines	Energy Regulatory Commission	Independent	Price pool
Singapore	Energy Market Authority	Under the Ministry of Trade and Industry	Price pool
Thailand	Energy Regulatory Commission	Independent	Single buyer
Vietnam	Electricity Regulatory Authority	Under the Ministry of Industry	Cost-based pool

Source: Adapted from Syaiful (2015)

Market design

Market design is a broad concept covering several areas such as allocation of costs, revenues, and rights to use the cross-border transmission lines; transmission pricing and allocation; balancing and congestion management; market clearing and dispatch rules in an interconnected market; and other regulations on renewables and taxation. The market design should be able to produce correct price signals for investment in both power generation and cross-border transmission infrastructure. Such endeavours are considered challenging, even in mature markets such as Nord Pool. These issues are very difficult to resolve without establishing institutions with the specific objective of promoting cooperation between TSOS and regulators. Such community initiatives are clearly lacking under the ASEAN framework.

Case study: Greater Mekong Sub-region (GMS)

The GMS is a natural economic area in the Mekong River Basin, consisting of a geographic area of 2.6 million square kilometres and a population of around 326 million. The area is well-endowed with various energy resources which have been exploited to different extents. The GMS power interconnection is the most significant scheme of bilateral power trade agreements in ASEAN. Although it has not yet moved beyond the initial phase, the GMS scheme provides relevant expertise for future projects in ASEAN to compare and learn from as it develops. It is particularly relevant to analyse the involvement of Lao PDR and Thailand, which are also involved in the LTMS concept.

Motivation and context for trade

While the potential for power exchange in the GMS region had been evident for some time, no significant efforts had been made to realize them until recently. The fast expanding economies in the region, particularly Thailand, and the conclusion of regional conflicts reinvigorated the impetus for regional power trade between GMS countries.

While there are shared interests, underlying motivations remains largely diverse. In lieu of its rapidly growing electricity demand (12% per annum growth rates), Vietnam is increasingly interested in importing electricity from neighbouring countries to supplement domestic generation capacities. Cambodia, on the other hand, views the potential to replace its high-cost and polluting oil-fired generation fleet with cheap and green hydropower from its neighbours to be attractive. Moreover, with its significant hydropower reserves, Cambodia may benefit from cross-border investment, which is essential for Cambodia to exploit its hydropower potential and become a net exporter in the future. China and Lao PDR also stand to benefit from increased foreign reserves through exporting electricity to their neighbours.

Recognizing the regional benefits of interconnection, various international donors, such as the Asian Development Bank (ADB), World Bank, and the Energy Sector Management Assistance Program, have supported assessment projects aimed at quantifying the potential benefits of cross-border electricity trade. It is estimated that a fully integrated electricity market in the GMS region would result in a 15% reduction in electricity costs, equivalent to US$213 billion, from 2005 to 2025. The cost savings will arise due to the following reasons:

- Cross-border interconnections can be more cost efficient than connecting the locality via the national grid from domestic plants far away. Proof of this are the many low voltage cross-border connections for the purpose of rural electrification. On a larger scale, it is also cheaper to generate and send power from southern Lao PDR into southern Vietnam than to do it from northern Vietnam.
- Due to advantages in natural resources, some countries can produce electricity easier than others. One clear example is the geographic endowment of Lao PDR, enabling it to tap into its massive hydro reserve, exporting power to Thailand, Myanmar, and Yunnan Province in the People's Republic of China (PRC).
- Using differences in peak load profile, it is possible to smooth out the overall load on the regional grid between Vietnam, Thailand, Myanmar, and Yunnan Province in the PRC. Further, using seasonal peak load variations, it is possible to balance out differences between Thailand and Vietnam during winter. The trading of electricity between areas balancing out load centres can help better utilize existing plants and postpone the construction of new generating capacity.

However, an estimated US$585 billion of investment is required in infrastructure to support the economic growth and the corresponding growth in electricity demand in the region (ADB 2013).

Historical development

Prior to the 1990s, only Thailand and Lao PDR possessed interconnection links. However, since the stabilization of the political situation in the GMS in the early 1990s, significant infrastructure development has been undertaken, with initial funding for the construction of interconnectors being supplemented with private sector financing and ADB support (ADB 2013). In those times, collaboration focused on building trust and frameworks between countries through different studies and consultations. ADB published the first GMS energy study in 1994. The report highlighted both the immense potential of, and mechanisms to facilitate energy cooperation in, the region. Recognizing the common benefits to be reaped, the report led to the formation of the annual Electric Power Forum in April 1995, which in turn facilitated the instalment of a formal policy statement promoting regional power trade initiatives in 1999. Eventually, the collaborative climate between parties in the GMS led to an intergovernmental agreement to implement the policy statement.

In 2000, the Regional Indicative Master Plan on Power Connection was initiated with the help of ADB. Completed in 2002, this document contained projected regional electricity demand in 2020 and identified key points for interconnection between the participants. In 2004, the GMS governments signed an intergovernmental agreement establishing the Regional Power Trade and Coordinating Committee (RPTCC). The RPTCC contributed significantly to the integration of the GMS region with its role in facilitating the completion of the initial Regional Power Trade Operating Agreement, which set forth the technical and commercial guidelines for a regional power market in the GMS. The MoU was signed in 2007.

Stages of market integration

Owing to differences in technical and regulatory standards between GMS participants, the MoU contained the roadmap of the market development in the four stages outlined next:

- **Stage 1 (current stage).** In Stage 1, power trade is almost entirely conducted under bilateral agreements between the PPA member (which can be national utilities or IPPs) selling power to the national utility of the importing country. Opportunities also exist for national utilities to trade power should there be excess cross-border transmission capacity.
- **Stage 2.** This stage foresees the presence of transit trade, wherein a pair of GMS countries may take advantage of the transmission capacity of a third member country to trade electricity between themselves. However, such trades must be subject to available transmission capacity in excess of that linked to PPAs.
- **Stage 3.** This stage entails further development of interconnectors and granting trade access to third parties other than national power utilities.
- **Stage 4.** This stage stablishes a regional competitive power pool market with multiple sellers and buyers within and across the member

countries. A precondition for this would be the establishment of national competitive markets.

The second MoU was signed at the Third GMS Summit in March 2008 in Lao PDR. The purpose of the second MoU was to clarify the timelines up to 2012 in preparation for Stage 2. The associated key activities were defined in a roadmap entitled "Indicative Regional Master Plan on Power Interconnection in the GMS", published in 2010.

To establish a more permanent institution to coordinate the trade of power within the region, GMS members agreed on reforming the RPTCC into the Regional Power Coordinating Center (RPCC) in 2012. RPCC's responsibilities include monitoring and coordinating national transmission system operators, utilities, and cross border trades to ensure harmonization of market procedures and planning of interconnector infrastructure. Financing would be provided by contributions from member states. However, as of early 2018, there was still no consensus about which country should host the RPCC.

GMS participants

The GMS has six member countries: Thailand, Cambodia, Lao PDR, Vietnam, Myanmar, and the PRC. China is represented by two provinces: Yunnan and Guangxi. The energy resources in the region are heavily dominated by hydropower.

Power sectors of the member states remain dominated by state-owned utilities, although there are large numbers of IPPs in all countries except Vietnam. Cambodia, Lao PDR, and Vietnam have vertically integrated generation, transmission, and distribution infrastructure, while Thailand bundles its generation and transmission. In Myanmar, the transmission and distribution sectors are integrated. Vietnam has launched a pilot initiative to develop a competitive electricity market, but the development has been slow so far (ADB 2016). The current members who signed the MoU are Electricité du Cambodge (EDC), Electricité du Lao PDR (EDL), Electricity Generating Authority of Thailand (EGAT), Electricity of Vietnam, China Southern Power Grid (CSG), and the Government of Myanmar, which controls the power sector domestically. NRAs are present in Cambodia, China, Thailand, and Vietnam, albeit with varying degrees of independence and control.

It is important to note that some of the IPPs involved in cross-border trade are partially owned by the same state-owned enterprise which is purchasing and importing the power into the country. This is particularly the case in Lao PDR, where plants exporting power exclusively into Thailand are partially owned by EGAT.

Cross-border power trade

In capacity terms, the important interconnection lines in the GMS region are two lines between Lao PDR and Thailand with 1,500 MW and 1,000 MW transmission capacity, as well as one 600 MW line from Myanmar to China. The total interconnection capacity in the GMS in 2016 was approximately 5,000 MW, with almost 75% of capacity serving the link between Lao PDR and Thailand. In 2014, electricity imports across the GMS were estimated at 19.7 TWh, of which Thailand accounted for 12.3 TWh, having stabilized around the 20 TWh level for the preceding four years. Thus about 6.7% of Thailand's annual electricity consumption, which was 169 TWh in 2014, was imported (GMS 2017).

The breakdown of electricity trade data is shown in Table 5.8. The data are from the year 2013, which is the most recent complete data source available. The largest percentage of trade activity is by Thailand, accounting for almost 50% of total trade. This is intuitive considering the distribution of interconnection lines. China and Vietnam are the next largest trade participants, while the shares of Lao PDR and Cambodia are several times smaller. The total volume of trade in GMS has grown significantly since 2000, but appears to have plateaued after 2010. In a broad sense, Thailand and Vietnam are the most motivated to import power from the GMS market to meet their growing electricity demand. Thailand, in particular, is incentivized by the potential to replace expensive liquefied natural gas (LNG) imports with cheap hydro-based electricity from the GMS.

Electricity prices

Currently, electricity prices for bilateral cross-border trade are negotiated on a per contract basis. Some of the associated bilateral contracts and trade

Table 5.8 GMS Power Trade and Imports in 2013 (TWh)

Country	*Imports*	*Exports*	*Total Trade*	*Net Imports*
Cambodia	2.1	–	2.1	2.1
Lao PDR	1.2	1.1	3.3	−0.1
Myanmar	–	–	0	0
Thailand	12.6	1.3	13.9	11.3
Vietnam	3.7	1.3	5.0	2.4
People's Republic of China	1.9	3.9	5.8	−2.0
Total	**21.4**	**7.5**	**28.9**	**13.9**

Notes: – = nil. All values in TWh = terrawatt-hour.

Source: GMS (2017)

terms can be obtained from World Bank (2010). Overall, there is little publicly available information about power prices associated with each PPA. The available prices varied between USc5 and USc8 per kWh, in 2010 price terms, which are considered lower than the average electricity tariffs in importing countries. It is likely that the contracts have been re-negotiated, upwards, during the period of high oil prices and while countries like Thailand are progressively rolling back domestic energy subsidies.

State of supply and demand

No recent forecasts for the supply and demand situation are available for the GMS region as a whole. ADB reports that preliminary examination of energy demand forecasts were conducted by the ASEAN Center for Energy (ACE) in 2011, with the assistance of the Institute of Energy Economics, Japan, and expert teams from each of the member countries (ADB 2013). However, this study does not include energy forecasts for Yunnan and Guangxi, which are part of the GMS. The more recent Fourth ASEAN Energy Outlook does not provide disaggregated information for the member states but only for the region as a whole (IEA 2015). The report by the World Bank, published in 2010, estimated potential capacity additions in GMS countries derived from the official documents (Figure 5.2). Given that these data were collected more than eight years ago, these values should serve only as a rough estimate. Nevertheless, one can observe the significant anticipated growth in generating capacity which comes mainly from hydropower, coal, and gas power.

In addition, ADB (2013) has published an estimate of peak load demand for the GMS region, which can serve as a proxy for projected demand. Notably, due to economic and population growth, peak electricity demand is expected to increase more than threefold between 2010 and 2025, reaching 277 MW. The most significant growth is anticipated in Vietnam, Guangxi and Yunnan in the PRC, and Thailand.

Institutional arrangements

The GMS does not have a formal secretariat or developed regional institutions. Instead, emphasis is placed on the government agencies and national utilities of participating countries. As such, the ADB performs various administrative support functions, wherein it provides secretarial (coordinate meetings), technical, financial, administrative, and logistical support to the GMS institutions. Development of interconnection facilities does not follow any comprehensive regional planning initiative, but is a part of the long-term power development planning of all GMS member countries. The

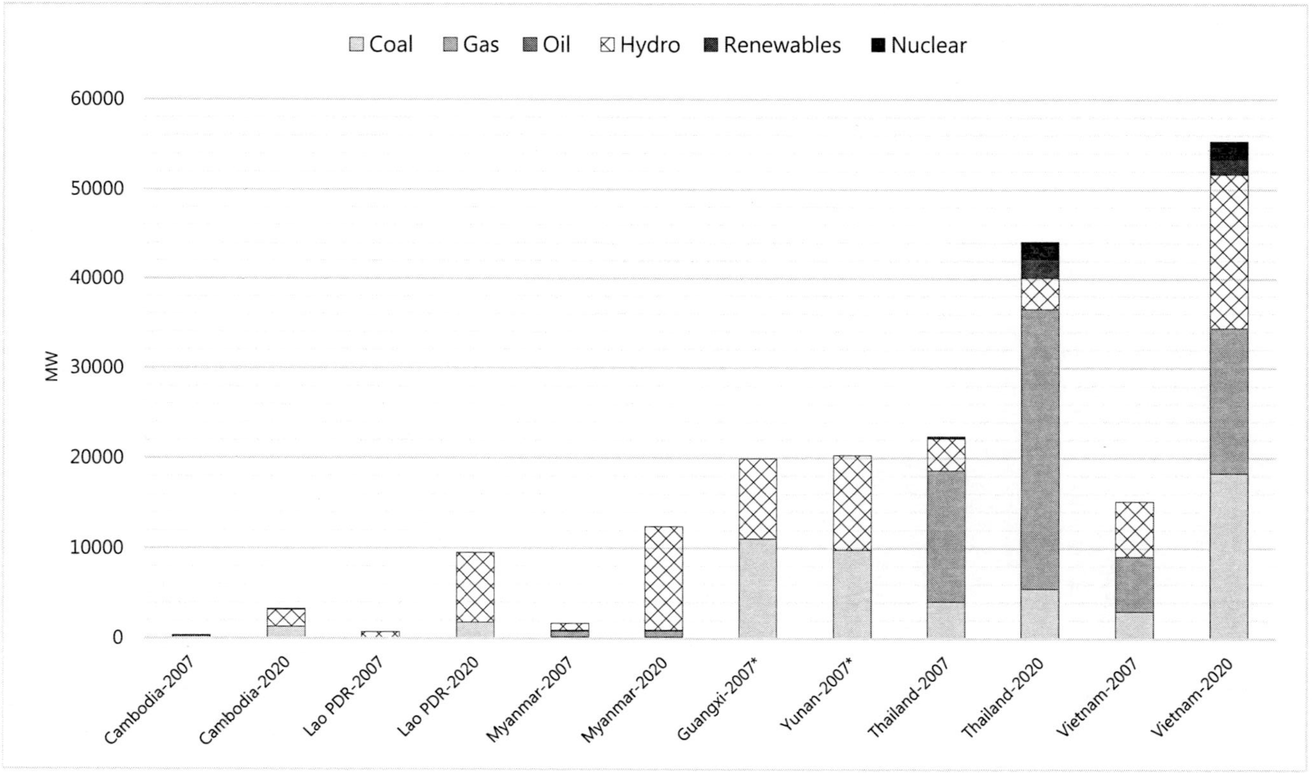

Figure 5.2 Comparison of Installed Capacity in GMS Region in 2007 and in 2020 (Projected)

Note: 2020 projections are not available for Guangxi and Yunnan.

Source: World Bank (2010)

AIMS provide details on prioritized projects; however, the member countries are not obliged to follow them.

Interconnector facilities are typically owned by the domestic transmission operators in respective member states. The case for Lao PDR is unique in that infrastructure is typically owned by the project development company. Funding for interconnection projects is raised through a variety of channels, including the national utilities, international banks, commercial banks, and donor/development agencies.

Despite existing bilateral power supply agreements between the countries, there is no overarching regulator across the GMS. Due to the different national circumstances of each country, national regulators have also different levels of independence from their governments. However, it should be noted that none can be described as fully independent and empowered. Lao PDR and Myanmar have no national regulators. Some cooperation was achieved between regulators in Thailand and Vietnam, who have stated the need to harmonize domestic grid codes to align with regional GMS performance standards (World Bank 2010).

Conclusions

The IEA has reported that, according to ADB estimates, the GMS power interconnection has resulted in US$14.3 billion in savings, coming mainly from the substitution of fossil fuel generation with hydropower (IEA 2015). ADB notes that despite some progress in efforts towards regional power market integration, there remains no material departure from the initial status since the signing of the Intergovernmental Agreement in 2002 (ADB 2013). The main obstacles for the GMS market to transition to Stage 2 are different technical standards and regulatory arrangements among the member states. Moreover, without some extent of reforms in national electricity markets, it is difficult to foresee national utilities coming to consensus through an intergovernmental MoU. The lack of adequate transmission network across the whole GMS region is another important constraint. Furthermore, it is unclear who will be financing new transmission lines given the asymmetric distribution of benefits and sharply varying levels of economic affluence in the area.

Ownership of dedicated cross-border transmission lines by IPPs creates additional challenges since third-party access is restricted. Currently, many major cross-border transmission lines are privately owned by the IPPs, who are unwilling to allow third-party electricity flow through their lines. In other cases, technical restrictions limit the potential to connect sources beyond the infrastructure dedicated to existing PPAs. As such, massive restructuring, if not a complete overhaul, of pre-existing PPAs is required to open up transmission capacity and introduce public-private partnerships into an integrated regional electricity market.

The issues are further exacerbated by growing geopolitical tension between ASEAN member countries and the PRC due to various territorial disputes and the use of water in the region. Hydropower development undertaken by China in the upper Mekong River region can restrict the flow of water downstream to Vietnam and Lao PDR, which is critically needed to generate and sell power there. Any kind of instability in one member country can also have adverse impacts on neighbours who are dependent on either importing the energy for power or exporting the electricity for foreign currency. A more detailed list of challenges and opportunities of the GMS can be found in ADB (2013).

References

ADB (Asian Development Bank). 2013. *Assessment of the Greater Mekong Sub Region Energy Sector Development: Progress, Prospects, and Regional Investment Priorities*. Manila: ADB.

ADB (Asian Development Bank). 2016. *REG: Greater Mekong Sub Region. Summary of Proceedings: 19th Meeting of the GMS Regional Power Trade Coordination Committee (RPTCC-19)*. Accessible at: www.adb.org/sites/default/files/publication/178869/rptcc-19-summary-proceedings.pdf.

AEMI (ASEAN Energy Market Integration). 2016. *Energy Security and Connectivity: The Nordic and European Union Approaches. Proceedings of ASEAN Energy Market Integration (AEMI) Forum*. Singapore: Energy Studies Institute. Accessible at: http://esi.nus.edu.sg/docs/default-source/doc/proceedings-of-aemi-forum-2015.pdf.

AIMS (ASEAN Interconnection Master Plan Studies) Working Group. 2010. *ASEAN Interconnection Master Plan Study. Volume I. Jakarta: ASEAN*. Accessible at: www.asean.org/storage/images/resources/ASEAN%20Publication/2011%20-%20Master%20Plan%20of%20ASEAN%20Connectivity,%203rd%20Reprint%20(Jun).pdf.

ASEAN (Association of South East Asian Nations). 2017. *ASEAN Statistics Leaflet – Selected Key Indicators 2017*. Jakarta: ASEAN. Accessible at: www.aseanstats.org/wp-content/uploads/2017/11/ASEAN-Statistical-Leaflet-2017_Final.pdf.

GMS (Greater Mekong Subregion). 2017. *Statistics on Growth, Energy, and Bilateral Trade in the Greater Mekong Subregion, Third Edition*. Accessible at: www.greatermekong.org/sites/default/files/gms-statistics-3rd-ed-view.pdf.

IEA (International Energy Agency). 2015. *Development Prospects of the ASEAN Power Sector*. Paris: IEA.

IEA (International Energy Agency). 2017. *World Energy Outlook Special Report 2017: Southeast Asia Energy Outlook*. Paris: IEA.

Roesli, A. P. 2006. *HAPUA and the ASEAN Power Grid for Optimum Use of Energy Resources', the Regional Energy Policy and Planning in ASEAN for Sustainable Development (REPP-ASD)*. Bangkok, Thailand.

Shi, X., and C. Malik. 2013. *Assessment of ASEAN Energy Cooperation Within the ASEAN Economic Community*, ERIA Discussion Paper Series 37. Jakarta: ERIA. Accessible at: www.eria.org/ERIA-DP-2013-37.pdf.

Syaiful, I. 2015. *ASEAN Power Market Integration*, presented at ACE-HAPUA-IEA-World Bank Workshop, 13 March 2015.

World Bank. 2010. *Regional Power Sector Integration: Lessons From Global Case Studies and a Literature Review (Vol. 4): Greater Mekong Subregion (GMS) Transmission and Trading Case Study (English)*. Energy Sector Management Assistance Program (ESMAP); Brief note 004/10. Washington, DC: World Bank. Accessible at: http://documents.worldbank.org/curated/en/793791468151161928/Greater-Mekong-Subregion-GMS-transmission-and-trading-case-study.

6 Lao PDR-Thailand-Malaysia-Singapore interconnection

Introduction

The Lao PDR-Thailand-Malaysia-Singapore (LTMS) interconnection is an initiative to establish the first multilateral power interconnection project in ASEAN. This project was initiated during a Special Senior Official Meeting on Energy, convened on 9–11 December 2013 in Manado, Indonesia. In September 2014, the countries agreed on the pilot project to study cross-border power trade from Lao PDR to Singapore via the existing interconnections between the transit countries. The LTMS interconnection would take into account each country's national development plan and existing interconnection among the countries, as well as associated laws and regulations (ACE 2015).[1]

The member states set up a working group to study the technical feasibility of a pilot project of up to 100 MW from Lao PDR to Singapore using existing interconnections. In this pilot agreement, Thailand and Malaysia would serve as the transit countries. The working group consisted of representatives from utilities and regulators and government agencies of the four involved countries. The task of the group was to set up a conceptual framework for the project by examining relevant technical, regulatory, political, and commercial issues that needed to be resolved before trade could commence.

In September 2016, the MoU was signed between Lao PDR, Thailand, and Malaysia, but Singapore did not sign the document. Under the agreement, Malaysia would buy up to 100 MW of power from Lao PDR, to be transmitted through Thailand's national power grid. The project could potentially be extended to Singapore in the future. No reasons were given why Singapore chose not to participate in the MoU, although it should be noted that the country is very different from the other project participants in terms of energy landscape and market structure.

The case for the LTMS project

The LTMS project has been identified by ERIA as the most promising multilateral power trade scheme connecting Lao PDR to Singapore. The associated economic net benefit could be US$25.5 billion (ERIA 2015). However, the scope of impacts caused by the interconnection does not include environmental and socioeconomic risks. In particular, cost assessments failed to consider the vulnerability of the ecosystem of the Mekong River Basin to the construction of associated hydropower dams and new transmission lines that would be necessary for growing power exports from Lao PDR. To date, the LTMS pilot project does not foresee any new infrastructure investments, as it caters only up to the existing 100 MW of capacity between the countries.

Lao PDR has approximately 18 GW of hydropower potential, which can be utilized as a relatively easy source of foreign currency revenue for the country. However, Lao PDR also carries a large burden of potential environmental damage that has not been fully accounted for in the cost-benefit studies. Thailand and Malaysia have the largest peak demand and annual consumption in the region, and are, therefore, particularly interested in hydropower imports from Lao PDR. Considering that electricity demand in Thailand and Malaysia is projected to rapidly increase over the next couple of decades, Thailand's motivation is to fuel this growth without relying heavily on high-cost natural gas imports. Malaysia has a somewhat similar motivation, although it has advantages compared to Thailand in that it has larger domestic reserves of oil and gas, and an alternative of potential electricity imports from Sarawak.

For Singapore, the motivation is a combination of benefits and drawbacks. The country has currently a 40% overcapacity in the generation sector and low projected electricity demand growth. Importing hydropower into the domestic market, which is almost entirely natural gas-based, will put further downward pressure on the wholesale price. This benefits larger consumers (i.e. industry), but negatively impacts on the domestic power generation companies. While hydropower imports would enhance energy security benefits by reducing Singapore's dependence on natural gas imports, conversely it may also reduce it by making Singapore dependent upon imports crossing three countries and subject to the vagaries of regional politics.

Lessons from international experience

The development of the LTMS project resembles, in many ways, other examples of energy cooperation in ASEAN, such as the APG and TAGP. Paved with many negotiations and delays, the actual progress is moving ahead slowly. This is mainly due to the numerous barriers for cross-border power trade in ASEAN, discussed in Chapter 5, which need to be resolved before actual power exchanges can take place. In this section, we compare the situation in the LTMS (and potentially other regional interconnection initiatives) with the situation in the SAPP, the European market, and Nord Pool at similar stages in their development, and extrapolate relevant experience from those markets for the LTMS project. Specifically, we draw on some common themes from the various energy market integration components, which are, broadly, (1) provision of physical infrastructure, (2) market design and operations, and (3) institutional arrangements. We shall also consider the additional issues of harmonizing national regulators and environmental, social, and security concerns.

Physical infrastructure

New capacity development

Currently, the physical interconnection between the LTMS countries has a maximum capacity of 300 MW, which is more than enough to support the LTMS pilot project of 100 MW capacity. However, additional facilities will be needed if the level of integration between the countries is planned to increase. This includes generation transmission lines that either connect existing infrastructure networks in neighbouring countries or create new networks across countries. Coordination of infrastructure investments is an important task between LTMS countries, who can draw on experience from Europe and Nord Pool. In Europe, ENTSO-E and ACER prepare TYNDPs, which represent a coordinated planning initiative for pan-European transmission that is agreed by all European TSOs and regulators. The Nordic countries did not have an institutional counterpart to ENTSO-E; this task was carried out by the planning committee of Nordel,[2] made up of representatives of Nordic's TSOs.

Ideally, network development plans for interconnected markets should be based on similar planning horizons across all markets. An added complexity is to incorporate into the demand forecasts all national and regional policy targets to ensure a uniform and coordinated approach to the planning process. Finally, the plans should be aligned with regard to project financing and implementation timelines. Experience from the SAPP shows that developing a regional infrastructure development plan without considering these aspects may not lead to effective project implementation.

Cost-benefit analysis is the common tool when prioritizing new projects. However, project selection should not only be motivated by cost reduction but also take into account relevant factors, such as national security interests and environmental and socioeconomic factors, among others. The TYNDPs include environmental impact assessments, an aspect that should also be considered for the LTMS initiative, particularly in the Mekong region.

The extent to which TSOs should be involved in interconnection capacity development is an open question. Clearly, TSOs (and merchant interconnectors) are the main operators of transmission lines, and thus cannot be excluded from the planning exercise. However, experience from the TYNDPs has highlighted the possible conflict of interests between the TSOs, who benefit from the congestion rent while at the same time undertaking capacity planning and prioritization.

Operation of existing networks

The physical infrastructure between LTMS member states is developed enough to allow some basic power exchange, but it is neither synchronized nor does it operate according to a common standard. Although international power markets studied in this book illustrated that having a single system operator is not a pre-requisite for interconnection, power markets do require a harmonized set of rules on operation. These rules are known as the network codes and can be divided into (1) grid connection codes, (2) system operation codes, and (3) market codes. The first two categories deal with access to, and operations of, the physical system and comprise a range of topics, such as requirements for obtaining grid access, operation security, operational planning and scheduling load frequency control, and HVDC connections. These rules are legally binding for all network participants. Network codes are drawn up nationally or even sub-nationally in many countries. However, as the interconnected area grows in size and complexity, such arrangements must become supra-national to ensure smooth operations.

Experience from the SAPP and the GMS shows that cross-border power exchange can happen even without a comprehensive alignment of these standards. However, there is a need for these codes to be harmonized – i.e. the codes implemented in different countries should not interfere with each other. The absence of harmonized grid codes is viewed as a limiting factor for both the SAPP and the GMS, which also causes frequent technical breakdowns in both systems (IRENA 2013; ADB 2013). In Europe, network codes are developed by the EC, ENTSO-E, and ACER, which also cooperate with Nord Pool. Prior to that, Nordel's operational committee was responsible for developing these standards across all the Nordic states. LTMS countries could harmonize grid codes in stages, first bilaterally and then multilaterally.

Another challenge related to power systems is that some of the existing transmission lines, particularly between Lao PDR and Thailand, have been "locked" by PPA agreements with IPPs and therefore cannot be used to transfer power between entities other than the contractual parties. Renegotiating of these agreements may be required to obtain full access to transmission lines.

Financing

The financing arrangements for cross-border power infrastructure is another important issue. According to ERIA estimates, the cost of the proposed future interconnectors on the LTMS route, listed under AIMS II, would be US$2 billion (ERIA 2015). Project financing is typically undertaken by beneficiaries of the interconnectors, which are national utility companies, independent TSOs, and/or third-party investors. For example, in the SAPP, finance for interconnectors is either raised directly by the utilities involved or through an SPV set up to execute the projects. In Europe and in Nord Pool, the TSOs and third-party investors carry the bulk of the financial burden. The involvement of national governments and the EU is low, although it should be noted that it is easier to obtain public financing if the project is of common interest. Access to financing and favourable financing terms is a challenge in all three markets, but most notably in Southern Africa. Some of the key projects have been financed through international development agencies and donors due to urgent socioeconomic conditions and lack of viable alternatives. Financing for the GMS has been a mixture of public and private sources, with a lot of support from the ADB and the World Bank.

Financing for the LTMS will likely be a mixture of public and private channels, sourced from national utilities, IPPs, and third-party investors. Given the relatively high per capita income of the countries involved (other than Lao PDR) and the high expected benefits of the interconnection, obtaining direct funding from development agencies may be difficult. On the other hand, private investors typically come with a perception that the transmission business is risky and unattractive even in more mature markets (such as the EU), while small IPPs may not have enough funds available. Achieving stable returns and investor-friendly conditions and instruments are therefore paramount for access to sources other than domestic financing. Nevertheless, the involvement of dominant state-owned utilities in the LTMS and Singapore's strong credit rating could improve the perception of transmission system investments in this scheme. Credit guarantees from international development agencies could be another way to reduce default risk and attract investment.

Market design and operations

Choice of market design

There are a variety of models of cross-border power trade with different degrees of market integration (IEA 2015). A model with a low level of integration would be the "old" Nordic scheme, which relied on a system of bilateral contracts between state-owned utilities trading excess power. A similar model was adopted by the SAPP in its early years. Both markets have since adopted the multi-seller, multi-buyer spot market structure, albeit to a different degree of success. One of the important drivers for market reform was the mismatch between supply and generation capacities in the long run. Whereas excess supply, as occurred in the early 1990s in Norway, primarily affects power plant owners and investors, the impact is much more dramatic if there is a supply shortfall, as is the case in the SAPP today. The SAPP experience also shows that a market can initially have an oversupply problem that later turns into a supply gap when either power sector expansion cannot keep pace with the rapid growth in electricity demand, or supply shortfalls occur with hydro plants in times of drought.

While power exchanges or spot markets provide flexibility and competition benefits compared to a purely long-term contract market, they also cannot not fully address the issues of long-term capacity planning. As wholesale electricity prices became increasingly volatile, European and Nordic electricity sectors turned to various support policies and market mechanisms, such as feed-in-tariffs, contracts for difference, and capacity markets. These instruments aim to provide sufficient planning horizons for new investments. However, it is questionable that such advanced mechanisms can be implemented in the early stages of market integration. Given that ASEAN countries have a much higher demand for new capacity and investment than Europe, the choice of market design must address this issue at the initial stages.

The existence of a functional spot market does not necessarily mean that it will be prioritized for electricity trading. Two key lessons can be derived from the SAPP on this issue. First, in an interconnected system with frequent generation and transmission bottlenecks, parties prefer long-term contracts, allowing sides to agree on delivery terms in advance, as compared to trading in the spot market. Second, the spot market is an open trading platform which requires participants to disclose information about price, volume, and time of delivery to the market operator, regulator, and other participants. Bilateral trading does not have such requirements. Naturally, there is an incentive for existing players to continue trading via long-term contracts, especially if market concentration is high. The large difference in liquidity ratios of the spot markets in Europe implies the same problem. Nord Pool, which has the highest share of spot trading in Europe, tackled this problem by mandating that large utilities keep a certain share of trading in the spot market.

Choosing market design is particularly difficult for the LTMS project due to the existence of a competitive electricity market in Singapore. Foreign participants could be granted limited access to Singapore's market in principle, trading power on pre-arranged terms (e.g. by using vesting contracts). However, agreeing on such terms may be difficult, given that hydropower would undermine the competitiveness of the non-subsidized gas-fired generators in Singapore.

Reforming the existing power sector structures

An important question in the regional context is to what extent existing power sector structures are suitable for cross-border electricity trade. Among LTMS countries, Lao PDR, Thailand, and Malaysia operate under single-buyer models with IPPs, while Singapore has a liberalized power market with unbundled generation and transmission. Technically, none of the countries have structural barriers in place that would prevent cross-border power trade. The scheme can be established between Lao PDR, Thailand, and Malaysia linked by multilateral agreements, similarly to the SAPP. The Nordic market prior to 1990 could be another example, although the Nordic states were historically more competitive compared with the SAPP.

Including Singapore in the LTMS arrangement would be more challenging, because the power sector structure is more advanced in Singapore. Singapore allows only private generation companies to participate in generation and trading of electricity, while the government's participation is not permitted due to its exclusive transmission system rights. Therefore, Singapore does not have an institutional counterpart to join the inter-utility agreement in the LTMS.

Another important question is whether unbundling and privatization of single-buyer utilities is important for cross-border power markets to evolve further. The SAPP achieved a basic cross-border power trade without unbundling or privatization, albeit that the former reform has been recommended for the SAPP. Nord Pool became a successful power market with unbundling, but no privatization. Many European power markets have experienced both successful unbundling and privatization.

In general, unbundling is an important priority for market development, although the privatization process could be delayed. In cross-border power markets, unbundling will give all generators equal access to transmission systems and ensure that the single-buyer utility cannot easily abuse its transmission rights to prevent third parties from network access. Unbundling would also prevent cross-subsidies among generation, transmission, and distribution arms owned by the same entity, which would also raise the level of trust in the market and improve its liquidity. It may be realized as unbundling of accounts, legal unbundling, and ownership unbundling. Depending on the level of transparency and trust, not all unbundling forms may be required. Currently, only Singapore has successfully adopted unbundling of generation and transmission among LTMS member states.

Harmonization of market codes

Market codes is a sub-category of network codes which deal with electricity market rules. The codes govern spot market trading, transmission capacity allocation, definition of bidding zones, and provision of balancing services. In general, the same rules apply here as mentioned for other network codes: the codes do not have to be identical, but they must be harmonized to the extent that supports smooth market operations. Most electricity markets contain a set of market rules and codes, some of which are mandatory, while some can be negotiated between operating members.

Some market codes will have to be created specifically for the LTMS scheme, such as the definition of bidding zones or nodes, allocation and pricing of interconnection capacity, and provision of balancing services. The proposed mechanism under the LTMS pilot project adopts a MW-km transit charge for pricing of transmission capacity, which is similar to the method in the SAPP. Balancing would be provided by national system operators, while the cost of balancing is contained within the wheeling charge. This model may serve well for the initial phase of market operations with limited cross-border trading. However, the expansion of the pool may require adding transmission lines between countries and more efficient methods of transmission capacity allocation.

Establishing a market operator

Markets in advanced stages of multilateral cross-border trading involving power exchanges require an independent market operator. The operator would be in charge of the monitoring and management of electricity trade between countries and would act as a platform for connecting buyers and sellers. It would also provide inputs for national system operators for allocation of dispatch orders, physical transmission capacity, and provision of balancing services. Market operators have been established in all studied international power markets. Among LTMS countries, only Singapore has an independent power market operator, which is the Energy Market

Company (EMC). Although the EMC operates solely within Singapore's borders, it is theoretically possible to extend its role to Lao PDR, Thailand, and Malaysia, although this would be a technically challenging task. An important issue here is to ensure that the EMC would be seen as a truly independent market operator by countries other than Singapore. This could be achieved by strict surveillance and disclosure practices that became prevalent in both Europe and Nord Pool.

Institutional arrangements

The basic model for cross-border power trade can function on a purely bilateral basis, as was shown by the Nordic countries in the 1970s and 1980s. Relevant negotiations, harmonization of standards, and market operations could be carried out jointly by the state-owned utilities of the participating countries. However, more advanced stages of market integration naturally require inter-regional institutions driving integrated planning, harmonization of codes, and coordination of development efforts among many participants. The Nordic countries delegated these functions to various committees of Nordel, while cooperation of regulators is realized under NordREG. The SAPP has a similar arrangement with the SAPP committees and RERA. In Europe, these functions are carried out by separate institutions which are the EC, ENTSO-E, and ACER. Considering that the LTMS and ASEAN countries envisage full market integration at some point in time, there will be a need for regionwide institution(s) with additional powers and responsibilities.

An important question in this context is how much authority should be given to such institutions. The EU member states collaboratively developed new standards, but only became bound by them once they were called into force by the EC. In Scandinavia, Nordel holds an advisory role wherein their recommendations guide decisions on technical regulations. However, from these two examples it is difficult to find much relevance for ASEAN countries, given that geopolitical and socioeconomic conditions are vastly different in ASEAN. The lessons from the SAPP provide a more immediate example, wherein prioritization of the national interests over regional goals and lack of empowered inter-regional bodies curbed the progress of market integration.

Even though the ASEAN community has already established some inter-regional institutions related to the development of the APG,[3] none of them has binding authority. Furthermore, some experts argue that due to large differences in their national circumstances, it may not be practical for all ASEAN countries to come to a consensus on, much less to abide by, the same rules.

Given slow progress in the development of the APG, and the apparent withdrawal of Singapore from the LTMS MoU, we suggest first creating an institution with a limited intervention authority that serves as an information exchange and a capacity building platform. This would allow member states to build more trust in the APG without compromising national interests. Over time, this institution should ideally evolve into a more empowered functional body that can produce binding regulations and drive project implementation.

Harmonizing national regulations

In interconnected markets, national regulators have the challenging task of balancing national priorities and the requirements of inter-regional collaboration. Power sector regulations do not have to be identical across interconnected markets, but they should be harmonized to the extent whereby it prevents them from conflicting with developments in other relevant countries. Among LTMS countries, there are large differences between Singapore and the other three countries due to inherent differences between a competitive market and a single-buyer regime. For example, the independence of national regulators in Lao PDR, Thailand, and Malaysia could be questioned, because public sectors in all three countries are involved in generation, transmission, and regulatory functions. In Singapore, the public sector is involved only in transmission and regulatory functions. At the same time, no domestic power company in Singapore is mandated to import power, either by itself or through foreign representatives, while the government is not permitted to participate in electricity trading.

Lao PDR, Thailand, and Malaysia have a variety of regulatory support schemes in the electricity sector such as direct producer subsidies, feed-in-tariffs, and tax breaks. All of these subsidies distort the real price of electricity when it is traded in a competitive market, undermining the competitiveness of non-subsidized participants such as power companies operating in Singapore.

Environmental, social, and security concerns

The role of hydropower potential in the Mekong Basin is crucial in the current vision of ASEAN electricity market integration. Countries in the region, and outside of it, see it as means to offset expensive fossil fuel use and cut greenhouse gas emissions. However, intensive damming required for electricity generation will create profound environmental and socioeconomic impacts for the region, many of which are not properly understood. The magnitude of these impacts will grow disproportionally as increased interconnection expands the opportunities to purchase hydroelectricity to an increasing number of sites and countries.

Virtually none of the regional studies prepared by experts in the energy field analyse the environmental and social impacts of increased hydropower development in the Mekong Basin. In contrast, the 2010 study by the International Centre for Environmental Management (ICEM) notes a range of potential consequences including the following (ICEM 2010):

- Lower Mekong Basin (LMB) mainstream hydropower would deliver very significant economic benefits for the regional power sector, most of which (70%) would fall to Lao PDR (US$3–4 billion per annum);
- The losses experienced by the fisheries and agriculture sectors due to the mainstream dams are an order of magnitude greater than the realistic benefits to those sectors;
- The LMB mainstream projects would induce significant additional basin-wide effects on the Mekong River–dependent ecosystems, the majority of which are unavoidable if the projects go ahead;
- Even with mitigation measures conventionally associated with hydropower projects in the region, LMB mainstream projects would likely contribute to a growing inequality and a short- to medium-term worsening of LMB poverty in LMB countries;
- The mainstream projects would lead to irreversible losses in aquatic and terrestrial biodiversity of global importance;
- By 2030, if 11 mainstream dams were built, the protein at risk of being lost annually would be the equivalent of 110% the current annual livestock production of Cambodia and Lao PDR; and
- Risks and losses incurred by the Mekong terrestrial and aquatic ecosystems will result in increasing food insecurity for millions of people.

The conclusion of the report team was to defer any construction of mainstream dams for a period of ten years (ICEM 2010).

Notes

1 Regional motivation for power trade can be understood more easily upon reading the detailed review of participating countries in the annex of this book.
2 In 2009, this task was transferred to ENTSO-E when Nordel was wound up.
3 HAPUA (promoting of regional interconnection), ACE (project prioritization, harmonization of technical codes for APG), and AERN (alignment of regulatory and legal framework for APG).

References

ACE. 2015. *ASEAN Plan of Action for Energy Cooperation (APAEC) 2016–2025*. Jakarta: ACE. Accessible at: www.aseanenergy.org/wp-content/uploads/2015/12/HighRes-APAEC-online-version-final.pdf.

ADB (Asian Development Bank). 2013. *Assessment of the Greater Mekong Sub region Energy Sector Development*. Accessible at: www.adb.org/. . ./assessment-gms-subregion-energy-sector-development.pdf.

ERIA (Economic Research Institute for ASEAN and East Asia). 2015. *Study on Effective Power Infrastructure Investment Through Power Grid Interconnections in East Asia*. Jakarta: ERIA. Accessible at: www.eria.org/RPR-FY2014-30.pdf.

ICEM (International Centre for Environmental Management). 2010. *Strategic Environmental Assessment of Hydropower on the Mekong Mainstream: Final Report*. Prepared for the Mekong River Commission. Australia: ICEM. Accessible at: http://icem.com.au/documents/envassessment/mrc_sea_hp/SEA_Final_Report_Oct_2010.pdf.

IEA (International Energy Agency). 2015. *Development Prospects of the ASEAN Power Sector*. Paris: IEA.

IRENA (International Renewable Energy Agency). 2013. *Zambia: Renewable Energy Readiness Assessment*. Abu Dhabi: IRENA. Accessible at: https://www.irena.org/-/media/Files/IRENA/Agency/. . ./2013/RRA_Zambia.pdf.

7 Conclusions and the way forward

In the review of the experience of selected regional electricity markets around the world, we identified some key elements of integration that emerged independently as those markets evolved. These are coordinated physical infrastructure development, standardized or harmonized rules of operation, some form of market competition, and empowered governing or coordinating institutions. Prioritization of these elements and the sequence of steps to achieve them are not straightforward and depend on the regional market's environment and history. Moreover, in many regions, this process is still ongoing.

Market integration in Europe adopted a more top-down integration approach, capitalizing on the legal system of the EU. In contrast, Nordic and Southern African markets developed an incremental and voluntary basis, driven by the utilities themselves. Given diverse regional circumstances in ASEAN and the absence of an overarching legal system like in the EU, we believe that the latter approach is more suitable for ASEAN.

The importance of coordinated infrastructure development is particularly important in markets with growing electricity demand, such as the SAPP and ASEAN. Insufficient generation and transmission infrastructure in Southern Africa seriously limit the progress of the otherwise successful SAPP market. Frequent power outages and failures to deliver contractual obligations reduce the trust in the benefits of electricity market integration. Poor infrastructure development is driven by non-cost-reflective tariffs and weak protection of third-party investors. These aspects deserve consideration by ASEAN, where the required generation capacity is expected to double by 2040.

Another important issue is whether particular ASEAN countries should maintain a vertically integrated structure in the electricity industry in member countries. While this is matter for national policy in each sovereign country, we note that such structures do not preclude cross-border interconnection and power trade until a fairly advanced stage. Nevertheless, in order

Table 7.1 Three Market Design Options

Option 1 *Multilateral Trade of Excess Power via Long-Term Contracts*	*Option 2* *Multilateral Trade with Spot Exchange*	*Option 3* *Fully Competitive Power Markets*
Closest analogue:	***Closest analogue:***	***Closest analogue:***
Nordic countries before 1990	SAPP	Nord Pool, some European countries
Steps required:	***Steps required:***	***Steps required:***
– Formulation of institutional and contractual arrangements for cross-border power trade – Some harmonization of technical and regulatory standards – Coordination of system operation between countries for electricity transfers – Signing of contracts between state-owned utilities on pre-arranged terms – Setting up separate entity to trade power in country with competitive markets (e.g. Singapore) on pre-arranged terms – Transit charge is optional, although desired	– Formalizing the market institution with relevant committees – Setting up independent and empowered association of energy regulators – Agreement on coordinated infrastructure development plans – Development and adoption of comprehensive network codes, including grid connection codes, system operation codes, and market codes – Deeper harmonization of existing national standards with grid codes – Setting up market operator and legal market entity – Phasing out energy supply subsidies	– All steps under Option 2 plus: – Vertical unbundling of state-owned utilities – Full independence of TSOs from electricity production – Unrestricted and non-discriminatory grid access to all participants – High market transparency and access to information for all market players – Sophisticated methods of system balancing and transmission capacity allocation

Pros:	***Pros:***	***Pros:***
– Easy to implement – Does not require power sector reforms – Provides mutual benefits in system security	– More efficient than Option 1 – Provides greater benefits to all efficient participants – Can react to market signals – Creates pathway for Option 3	– Most efficient of all options – Reduces wholesale electricity prices – Provides greater benefits to all efficient participants – Increased market liquidity – Attracts private sector investments
Cons:	***Cons:***	***Cons:***
– Inefficient – Low flexibility as market signals are missing – Retains non-competitive practices – Infrastructure investments are difficult, particularly in transmission – Low private sector participation	– Presence of unbundled state-owned utilities deters private sector participation – Information asymmetry – TSOS are not independent	– Requires difficult domestic reforms – Requires high level of technical sophistication and experience in operating power markets – Requires stable political climate with high protection of participants' rights

to move the market to a more competitive structure, separation of generation and transmission industries is highly recommended.

Tapping on hydropower potential in the Mekong Basin is a crucial aspect of the current vision of ASEAN electricity market integration. We note that virtually all existing regional interconnection studies do not sufficiently analyse the profound environmental and socioeconomic impacts that come with the damming of Mekong and its tributaries. Potential impacts include (1) loss of biodiversity of global importance, (2) increased food insecurity for millions of people, and (3) increased geopolitical tensions that could outweigh the collaborative push resulting from electricity market integration. All of these aspects should be considered by policymakers in ASEAN when designing a vision for a common resource use.

Given the commitment of ASEAN member countries to increase crossborder interconnection and power trade, we suggest three market design options and indicate the required steps to achieve them. In setting out these options, we have sought to incorporate important lessons derived from the international experience analysed in this study (Table 7.1).

Annex: overview of LTMS participants

Lao PDR

Country profile

The Lao People's Democratic Republic (Lao PDR) is a landlocked, one-party socialist republic, which shares borders with Cambodia, China, Myanmar, Thailand, and Vietnam. The Mekong River, which accounts for a large portion of its border with Thailand, is an important source of hydro-power resources for the country, which aims to be the "battery of Southeast Asia" through exporting its hydroelectricity. Lao PDR is an authoritarian state with a one-party system, and the country operated under a planned socialist economy for 11 years following the overthrow of its monarchy in 1975. Bogged down by severe economic hardship and foreign exchange woes, Lao PDR introduced the New Economic Mechanism in 1986, which opened its doors for trade and market liberalization.

Currently, Lao PDR is considered to be one of the fastest-growing economies in the East Asia region and has successfully graduated from its least-developed country status to be classified as a lower-middle-income economy by the World Bank. Lao PDR's strong growth rate is primarily supported by its power and mining sectors, which were identified to be fundamental drivers of its national economy (World Bank 2016a).

Lao PDR is well-endowed with domestic energy resources, both fossil fuel–based and renewable sources. Although it has no proven oil and gas resources, and thus imports all its petrol and diesel needs, Lao PDR has sufficient lignite reserves for about 2 GW of installed capacity and 500 MW installed capacity for bituminous and anthracite coal. While the solar, wind, and geothermal potential for power generation is limited, biomass and hydropower are very attractive options. Traditional biomass (mainly in the form of wood fuel) remains a primary energy source, accounting for close to 70% of total energy consumption. In recent years, these traditional energy sources have been increasingly replaced by petroleum and electricity. Hydropower remains the most exploitable resource for Lao PDR. The total estimated hydroelectric generation potential stands at 18–20 GW, of which only about one-sixth was installed generating capacity as of 2017. Taking into account current hydropower under construction, it is expected that by 2020, 40% will be exploited, of which over 30% of the technically viable resources will be for export (ADB 2013). As of 2016, the major part of final energy consumption was from the residential sector. Keeping in line with the GDP growth estimates, it is expected that the commercial and industry demand for energy will increase accordingly.

Energy subsidies and other incentives

The pricing of some energy-related commodities is regulated in Lao PDR. The Ministry of Industry and Commerce regulates gasoline and diesel prices by region. However, regular revision of these prices may also indicate alignment with market price movements. Moreover, with a tight fiscal condition and various development needs, the relevant agencies may find it hard to maintain fiscal spending on energy subsidies. Despite speculation, due to the lack of transparency on price composition and pricing mechanisms, it remains difficult to accurately ascertain the presence or absence of energy subsidies.

Historically, Lao PDR has had major subsidies on electricity tariffs. However, over recent years, there has been a gradual alignment of tariffs and the real cost of electricity production and distribution. There has been a general increase in tariffs, although the rate is different for different sectors. The current level of tariffs suggests cross-subsidies between different consumer groups. Specifically, the residential sector, particularly those who consume the least electricity, is subsidized by the sectors who are better able to afford higher electricity tariffs, such as the embassies and international organizations.

Electricity sector outlook

The power sector is one of Lao PDR's strategic growth sectors. Currently, government directives for the power sector development centre on three main objectives:

- Increase the domestic electrification ratio to 90% by 2020 through the expansion of generation, transmission, and distribution networks, as well as through off-grid installations;
- Promote public-private partnership development in hydropower projects as a way of increasing export revenues as well as honouring export commitments with neighbouring countries; and

- Promote high-voltage (500 kV) grid development as a pathway to power systems integration in the GMS.

The total installed capacity in Lao in 2016 was approximately 6,000 MW, and the government expect it to reach 14,000 MW by 2020. The majority of the generated electricity is exported to other countries through a series of bilateral contracts, while domestic consumption remains low. With economic and population growth, it is forecasted that domestic electricity demand will face rapid growth over the coming decade. The growth in electricity demand could be traced back to two fundamental causes: increasing electrification of the population and rapid industrial growth in future years.

There has been remarkable growth in the national electrification rate in Lao PDR over the past couple of decades, with access to electricity increasing from 15.3% in 1990 to 87.1% in 2016 (World Bank 2017). Overall, electricity demand is expected to increase from 8.45 TWh in 2010 to 68.82 TWh in 2035, representing an overall annual growth rate of 8.8% throughout the period 2010 to 2035. The ADB, on the other hand, estimates a higher annual growth rate for domestic electricity demand of 15%–18% (ADB 2013).

Currently, Lao's Power Development Plan details two alternative resources to meet growing domestic demand: (1) expansion of the national utility's (Electricité du Lao PDR (EDL)) generation capacity and (2) support the development of both domestic and export-based IPPs. Despite EDL's plans to increase its generation capacity by close to 600 MW by 2016, a substantial portion of the increased demand would still have to be met by IPPs, in some of which EDL is investing (ADB 2013). The Laotian government is seen to be actively promoting the participation of IPPs in the electricity sector, particularly in the hydropower sector. The Electricity Law, first introduced in 1997, thereafter revised in 2008 and again in 2011, promotes private participation in the electricity sector by encouraging private sector investment through public-private partnership arrangements, or a fully private mode. As such most of the capacity developed in recent years is held by IPPs.

To support the government's ambitious rural electrification and economic development plans, there needs to be an accompanying increase in transmission and distribution capacities. As of 2013, rural villages not yet connected to the EDL grid were supplied by some 85 diesel based or small hydropower mini-grids maintained by provincial entities and medium-voltage connections to neighbouring grids. While rural electrification is likely to still require electricity imports due to the cost of importing electricity when compared to extensive grid extension, the need to meet strong industrial demand in the south demands high-quality grid transmission infrastructure (ADB 2013).

Prior to 2009, Lao PDR had four independent regional sub-grids – namely Northern Grid, Central Grid 1, Central Grid 2, and Southern Grid – with 115 kilovolt (kV) and lower-voltage lines and substations (ADB 2013). The four sub-grids were not interconnected, which gave rise to forced power shedding and poor quality electricity. There are ongoing plans to integrate regional grids into a single national grid to ensure high-quality power supply in the national grid. From 2009 to 2011, Central Grid 1 and 2 have been combined, and mainly serves the Vientiane Capital and Luang Prabang (ADB 2013). Going into the future, EDL has plans to integrate remaining sub-grids with high-voltage lines (115 kV, 230 kV, and 500 kV) by 2020 (World Bank 2016a).

Expansions in the transmission and distribution infrastructure is likely to be financed by EDL, with parallel participation of the private sector, following a pattern observed over the past decade. ADB estimates that EDL contributed 3,922 km out of the overall 5,500 km expansions in transmission infrastructure between 2011 and 2015 (ADB 2013).

At the same time, exports are expected to increase disproportionally in order to maintain the economic growth targets set by the government. Despite abundant hydropower resources, Lao PDR imports electricity from other countries. This is due to the requirement to support domestic demand during dry seasons to maintain grid stability. One key area of concern is the cross-subsidies between import and export prices for electricity.

The generation mix in Lao PDR is dominated by large hydropower, with minimal solar energy used in solar home systems in rural areas and biomass gasification in specific uses in the industrial sector.

Lao PDR also has a renewable energy target to increase the share of renewable energy to 30% in the total energy mix by 2025 (ADB 2013). The renewable target is to be met with three broad strategies: (1) increased share of bio-fuel in the transport sector, (2) the introduction of wind generation in the fuel mix, and (3) increased share of thermal energy from renewable sources.

Summarizing future developments of Lao PDR's electricity sector, we conclude as follows:

- **Domestic electricity demand will continue expanding rapidly, placing increasing stress on undersupplied domestic electricity markets.**
  With higher electrification rates and economic growth, particularly in the industrial sector, Lao PDR will experience a surge in electricity demand which may place stress on the domestic electricity market. With most capacity additions serving foreign markets, more emphasis

has to be placed on expanding power generation capabilities serving local markets. Moreover, the current situation of exporting electricity will most likely persist, due to the cost benefits associated with it and the lack of grid expansion, especially to remote areas.

- **Demand for high-quality transmission and distribution systems are required to accompany growth in the industry sector and ambitious rural electrification plans.**
  The government's rural electrification programme will in itself require substantial investment in distribution and transmission capacity. More significantly, the economic development plans in the mining sector are also expected to contribute to an increase in domestic electricity demand. Further, compounding the challenges is the need to transport electricity supply in the Northern and Southern Grids to the key demand centre in the Central Grid, a feat that requires not only expansion of the transmission capacity but also interconnection of the four separate subgrids (ADB 2013).
- **There should remain a focus on export-oriented hydropower projects that are facilitated by increasing bilateral interconnections.**
  With electricity exports providing strong support for the local economy, the dominance of export-oriented projects is expected to be maintained (at around 90% of new capacity). While the state-owned utility, EDL, is likely to invest in domestic capabilities, private-sector players are likely to remain more interested in export-oriented markets until the issue of domestic electricity price subsidies is resolved.
- **Renewables (excluding hydropower) will play an increasing role in Lao PDR's electricity market, although the overall impact will remain low.**
  With a stated emphasis on alternative renewable sources, such as wind and solar, it is expected that the generation capacities of these technologies will rise. However, the overall impact of these generation resources is likely to be limited, given that the wind generation projects that are currently being exploited are at a nascent stage, whereas solar generation is currently limited to solar home systems. This leaves the electricity generation mix vulnerable to seasonality and the need for interconnection to maintain stability.

Development and organization of the electricity sector

The Department of Electricity is responsible for the national energy policy, formulating key policies for the power sector in the areas of generation, transmission, distribution, and rural electrification, while also holding the role of the regulator of the electricity market in Lao PDR.

The electricity market operates under a single-buyer model, with EDL purchasing all the electricity and distributing it to end-users through various provincial branches. EDL is one of the three state-owned enterprises under the Ministry of Energy and Mines (MEM), under the Prime Minister's Office. Lao Holding State Enterprise (LHSE), an SPV that is intended to maintain the government's share in export-oriented independent power projects, and the Electrical Construction and Installation Company, the company tasked with the construction of EDL's transmission and distribution networks, are also under the MEM. However, it should be noted that EDL has "a wide operational scope", "despite its de jure agent status", thus potentially indicating a relative degree of autonomy in making commercial decisions in a market setting (IEA 2015).

EDL is a vertically state-owned integrated company that is responsible for a broad range of operations in the domestic power market, from generation to transmission and distribution. While at the moment EDL still owns some generation assets, it seems to be increasingly shedding those assets to its daughter company – EDL-Generation (EDL-Gen) Company. EDL-Gen is a public company that is listed on the Lao Securities Exchange, which raises private sector capital to purchase generation assets from EDL, as well as take over the equity holdings held by EDL in IPP projects.

To meet domestic electricity demand, EDL also undertakes off-taker contracts from generation capacity operated by IPPs. It is interesting to note that the projects typically adopt a build-operate-transfer or a build-operate-own-transfer mode, after which the project will be operated by LHSE.

EDL has various payment settlement mechanisms with the various independent power projects. The tariff for sale of electricity from IPPs to EDL is negotiated on a case-by-case basis, with the final tariff approved by MEM. For independent power projects with a mix of domestic and export capacities, the projects have bilateral PPAs with both the cross-border off takers and EDL. Recalling that EDL is also responsible for electricity imports and exports, it also has payment settlement mechanisms with the utilities in Thailand and Vietnam (ADB 2013; World Bank 2016a).

The distribution of electricity is undertaken by EDL and EDL-Gen, both of which are also tasked with the construction of transmission lines and substations, as well as operations and maintenance services to other power generation projects in Lao PDR. As already mentioned, over recent years, segments of electricity distribution have been undertaken by private sector companies. In order to achieve the 90% electrification target by 2020, provincial governments have engaged the private sector to expand local grid services, especially in remote and rural areas. These are typically build-operate-transfer projects, whereby the infrastructure is bought over by the government at investment cost plus a reasonable return at the end of five years of operation by the private player (World Bank 2016a). However, policy-wise, there remains a strong preference for public grid expansion due to the potential to leverage on subsidized electricity prices. As such, it is unlikely that public sector dominance in the electricity market will be diminished in the medium term.

The role of electricity exports

Over the past couple of decades, Lao PDR has become the major net exporter of electricity in ASEAN, with total exports far outstripping its domestic consumption. Cross-border hydropower sales have been a major source of revenue for the government and access to foreign credit but has also contributed to the environmental and social impacts in Lao PDR and countries downstream on the Mekong River. The aforementioned rapid acceleration in access to electricity was partially funded through foreign credit provided by Thailand and China in exchange for electricity exports.

Table A.1 gives details of electricity generation, supply (or consumption), imports, and exports for Lao PDR. Note that supply, exports, and imports do not sum to generation due to "losses".

The significant hydropower potential in the country has attracted interest from various international developers and investors, which was supported by the revised Electricity Law by the National Assembly introduced in 2008. The law facilitated private sector participation in export-centric hydropower development. However, the process is not without controversy. Firstly, the development of large hydropower projects has been shown to adversely impact the environment, as well as cause negative social impacts such as displacement of local communities (Ulrich 2015). While there is a national development plan to address the environmental sustainability of hydropower

Table A.1 Electricity Generation, Supply, Exports, and Imports for Lao PDR

Year	*Generation (GWh)*	*Domestic Supply (GWh)*	*Export (GWh)*	*Import (GWh)*
2001	3,653.7	710.3	2,871.4	183.8
2008	3,717.0	1,915.7	2,315.4	844.5
2011	12,968.9	2,555.7	6,646.5	1,209.7
2012	13,056.8	3,075.0	10,668.4	904.3
2015	16,302.1	4,238.6	11,548.7	2,049.8

Source: Adapted from UN (2017)

developments, with some success stories, application is patchy to say the least (Ulrich 2015). Secondly, the government has been criticized on the lack of transparency with regard to the awarding of concessions to hydropower developers (ADB 2013). In this context, corruption remains rampant in Lao PDR, with Transparency International ranking Lao PDR as one of the most corrupt countries in the world. This has created issues with foreign direct investment, particularly in its infrastructure, due to the government's inability to enforce rule of law and business contracts, among other factors.

Correspondingly, the government has acknowledged the need to manage the environmental and social impacts, as well as enhance the screening, negotiation, and tendering process for hydropower projects developers. The Water Resources and Environment Administration was established to be the primary agency for the environmental management, responsible for "certifying the environmental and social impact assessments for hydropower projects, as well as working on broader assessments such as cumulative impact assessments, integrated water resource management, and strategic environmental assessments" (World Bank 2016a). The Department of Energy Promotion and Development was set up to be the government's main communication channel with prospective hydropower developers and investors, with the added responsibility of negotiating key hydropower deals with relevant stakeholders. An additional layer of accountability was added with the Ministry of Planning and Investment, who eventually signs the project agreements with the developers and/or investors (World Bank 2016a).

Projections indicate that Lao PDR's electricity exports will increase in the future at a higher pace than domestic consumption. This will make Lao PDR the target of competing electricity demands from neighbouring countries which will increase their reliance on Lao PDR (at least in the northern region). The regional over-reliance on hydropower, which is subject to seasonal weather patterns, could create domestic and international tensions in the case of water shortage, as has been the case in Southern Africa and Latin America. Although Lao PDR has power cooperation agreements with neighbouring countries, such as Thailand and Vietnam, to meet domestic demand during hydroelectric-supply shortages in the dry season, it cannot give a full guarantee considering the rapid growth of regional electricity demand.

Malaysia

Country profile

Malaysia is a federative state with three federal territories and 13 provinces consisting of two regions, Peninsular Malaysia and Eastern Malaysia, separated by the South China Sea. Peninsular Malaysia is located on the Malay

Peninsula. Eastern Malaysia, consisting of two provinces, Sabah and Sarawak, is separated by some 1,000 km of the water area from the peninsular portion. In our analysis, we look only at the peninsular part of the country and henceforth refer to it with the term "Malaysia".

Malaysia is considered an upper-middle-income economy and possesses one of the highest GDP per capita ratios in the ASEAN region. The World Bank (2017) projects that Malaysian GDP will grow between 4% and 5% per annum till 2020. The population of Peninsular Malaysia was 26 million in 2017, comprising 81% of the country's total population.

The Malaysian economy has diversified beyond the traditional agriculture and resources sectors with rapid growth in the manufacturing and services industries over the past 20 years. Due to Malaysia's extensive domestic fossil fuel resources, which include large reserves of oil, gas, and coal, the energy sector remains a critical sector for the growth of the domestic economy, contributing almost 20% of the GDP. The proven oil reserves in Malaysia in 2017 totalled 3.6 BBL, second only to Vietnam in the ASEAN region. The proven gas reserves in Malaysia totalled 42 tcf (EIA 2017). The majority of Malaysia's oil and gas reserves are located in offshore fields. Malaysia has some coal resources located in Eastern Malaysia. However, the majority if it lies in environmentally sensitive areas and is hard to extract. Currently, Malaysia imports about 90% of its coal requirements.

Although Malaysia was historically an exporter of oil and gas, growing local demand and declining production turned Malaysia into a net oil importer in 2013. In the gas sector, Malaysia's imports of natural gas have grown over recent years alongside rising LNG exports (IAE 2015). Malaysia is the world's third-largest exporter of LNG (behind Qatar and Australia). However, despite significant potential for renewable energy sources, such as biomass, solar, and wind, development of these resources has been slow, mainly due to application quotas set by the government for its Fifth Fuel Policy (which will be described later in this section).

Energy subsidies

Malaysia has various energy and non-energy subsidies in place. Most of the energy subsidies fall on petroleum products, liquefied petroleum gas (LPG), natural gas, and electricity. Malaysia also has various tax and investment incentives targeted to motivate investment in oil and gas exploration, particularly in offshore fields (EIA 2017). Investment incentives also exist for energy efficiency and renewable energy projects.

Reduction of energy subsidies has been one of Malaysia's key policy directives in recent years. The government spent large amounts on fossil fuel subsidies during periods of soaring oil prices. Over the same period, the national oil conglomerate, PETRONAS, recorded a high amount of foregone revenue due to the widening gap between domestic and international fossil fuel prices. Combined with other factors, this led to a policy shift on energy subsidies in Malaysia commencing in 2011. Malaysia raised electricity and gas tariffs in June 2011 and tied electricity prices to fossil fuel price fluctuations. Players in the power and industry sectors have also been required to pay higher prices for natural gas procured from domestic producers since 2014. Furthermore, subsidies have been completely removed for specific types of gasoline and diesel. As of January 2016, the regulated price of piped gas was increased to RM 18.20/mmBtu from RM 16.70/mmBtu. However, this regulated price was still below the market price of Medium Fuel Oil at RM 25/mmBtu and involved the government in a 30% subsidy to the power sector (Energy Commission of Malaysia 2016).

Electricity sector outlook

Growing energy demand is the main challenge faced by Malaysia. Due to a combination of both economic and population growth, Malaysia experienced nearly an 80% growth in total primary energy demand from 2000 to 2013. In 2015, Malaysia had the third-highest energy demand in ASEAN, behind Indonesia and Thailand (IEA 2015). Electricity demand in Malaysia has grown at a higher rate than energy demand. Electricity demand and generation doubled between 2002 and 2012, landing at 140 GWh in 2014. The main consumers of electricity in 2014 were the industrial sector (46%), followed by commercial (33%) and residential (21%) sectors. The transportation and agriculture sectors contributed to less than 1% of total electricity demand. Electricity access is almost universal throughout the country.

Malaysia's electricity sector is heavily reliant on fossil fuels. The fuel mix consists of coal (48%), natural gas (42%), and large-scale hydro and renewables (10%). The total installed generating capacity in Peninsular Malaysia was about 23 GW in 2016. Over the past 30 years, the fuel mix has moved towards gas and coal, while the share of oil has decreased.

Studies by international agencies anticipate that primary energy and electricity demand in Malaysia will continue to grow at a rate between 2% and 3% per year over the next decade (IEA 2015; EIA 2016). Coal is expected to be the dominant fuel in Malaysia's electricity mix, followed by natural gas and large-scale hydropower.

The challenge of growing demand has been reflected in several policy directives seeking to provide means to sustain this growth. Policies such as the Fifth Fuel Policy, National Biofuels Policy, and National Renewable Energy Policy and Action Plan promote energy efficiency and use of renewable energy sources in order to decrease consumption of oil and gas and alleviate pressure on domestic producers.

In 2001, Malaysia introduced the Five Fuel Policy in their Eighth Malaysia Plan, which listed renewable energy sources as one of their key energy sources, alongside oil, gas, coal, and hydro. Fiscal inducements were used as investment incentives for corporations to take up renewable energy sources at the initial stage. Feed-in-tariffs for renewable energy producers were introduced in 2011. The tariffs are financed through the Renewable Energy Fund. The fund is financed through retail customers with monthly consumption above 300 kWh who pay an associated premium of 1.6% on their electricity bills. However, it remains unclear whether these schemes could support the fund in the long term, as renewable capacity will grow and more money will be required to support the feed-in-tariffs. Addressing the last point, the government has imposed quotas on capacities receiving feed-in-tariffs and was considering curbing feed-in-tariffs for solar in 2017. Malaysia's policies also promote energy efficiency and conservation. Households are offered rebates of 5%–7% on energy efficient appliances. In addition, minimum-energy performance standards for products have been implemented.

Summarizing the initiatives outlined in the aforementioned policy documents, we present three conclusions for the future development of the electricity sector in Peninsular Malaysia:

- **Coal will overtake gas as the dominant fuel for power generation.** Projections by the Malaysian authorities and international institutions indicate that coal-fired generation will cover a large part of future electricity demand. At the same time, the proportion of natural gas in the fuel mix will decrease. Coal has a cost-competitive advantage over natural gas, which will become even more pronounced with the gradual removal of fossil fuel subsidies. Some of the new coal-fired generation capacity will utilize the ultra-supercritical technology in order to achieve lower CO_2 emissions.
- **Growing renewables will still be a small market player.** The development of renewable energy will accelerate, although renewables will not play a pivotal role in the fuel mix for at least another 10–15 years. One of the limiting factors is the existing cap on renewable energy capacity and the unclear long-term policy direction on feed-in-tariffs. Nevertheless, the cost-competitiveness of renewables vis-à-vis gas will increase with declining costs and gradual removal of fossil fuel subsidies.
- **Inter-regional power transfer will play a more important role.** Malaysia is planning to increase its power interconnection capacity by establishing links between Peninsular Malaysia and Sarawak,

Peninsular Malaysia and Sumatra, and expanding the existing capacities in the north with Thailand and in the south with Singapore. Specifically, official sources mention "on the ASEAN front, the Lao PDR–Thailand–Malaysia–Singapore Power Integration Project which is at an advance negotiation stage, will be a significant milestone in the ASEAN agenda, paving the way for future interconnection between ASEAN countries towards realising an ASEAN power grid" (Energy Commission of Malaysia 2016). Meanwhile, the inter-regional power flow from Sarawak will be delayed until at least 2025.

Development and organization of the electricity sector

The current model in the Malaysian electricity sector is a single-buyer model with monopolies in the transmission and distribution sectors and some form of competition in the generation sector. Each of the Malaysian states, including Peninsular Malaysia, Sabah, and Sarawak, have their own dominant utility company acting as a monopolist. In Peninsular Malaysia, the Tenaga Nasional Berhad (TNB) owns the National Grid (transmission), the majority of distribution assets, and some generation capacity (ERIA 2015). Market regulation is done by the regulatory body, the Energy Commission, which was established in 2001 and regulates both electricity and gas industries. Energy policy is overseen by the Energy Section of the Economic Planning Unit, one of the organizations under the direct jurisdiction of the Prime Minister's Department, the country's decision making body with the highest authority.

Malaysia has traditionally been a complete vertically integrated monopoly in the electricity sector. Reforms have begun with the Electricity Supply Act passed in 1990. This Act permitted private investment to enter the generation business as a means to respond to rapidly increasing electricity demand. Shortly after that, the first electricity purchase agreement with an IPP was signed in 1993. By 2013, a total of 14 IPPs in Peninsular Malaysia were selling electricity to TNB via 21-year fixed price PPAs (Ngadiron et al. 2014).

Since the introduction of IPPs, Malaysia has been exploring the option of further liberalization in the electricity market through an Independent Grid System Operator (IGSO). The proposed role of this IGSO is to schedule dispatch of power plants in a liberalized electricity market. However, this plan has never been realized due to a conflict of interests of the proposed scheme with TNB. Little development in electricity sector reform occurred in the following ten years. Starting from 2012, some attempt at unbundling was introduced to the single-buyer model. Prior to that, TNB was responsible for long-term capacity planning, energy contracts, and dispatch schedules as one entity. As a next step in market evolution, it was proposed to separate operations within TNB's single-buyer unit by creating four independent entities: (1) Power and Resource Planning, (2) Power Contract Management, (3) Enterprise Management, and (4) Technical Advisory and Industry Development. The Power and Resource Planning unit would be responsible for ensuring both long-term and short-term stability of the electricity market in Malaysia. The unit oversees daily operations by ensuring quality load forecasts and day-ahead generation scheduling. Furthermore, the unit also ensures medium to long-term energy security through its capacity planning, outage planning, and other strategic planning initiatives, such as the management of gas supply contracts, monitoring of fuel mix, etc. The Power Contract Management unit is responsible for the negotiating and managing all contracts with regard to the electricity market, including but not limited to PPAs and cross-border power trade contracts. The Enterprise Management unit is tasked with developing and managing matters involving regulations and compliance in relation to the relevant legislation, codes, or private agreements. Finally, the Technical Advisory and Industry Development unit was created to provide the necessary commercial and technical expertise to guide the regulator in policy decisions.

Thailand

Country profile

Thailand is bordered by Lao PDR to its east and north and by Malaysia to its south. The country has an area of 513,000 km^2 and a population of about 70 million. The World Bank (2016b) has estimated Thailand's GDP per capita at US$5,930. Thailand is the fourth-largest and richest country in the ASEAN region. It is classified as an upper-middle-income economy, and its GDP is expected to grow between 2.5% and 3% annually to 2020.

Oil and gas reserves in Thailand are in steady decline, while domestic energy consumption increases. Gas reserves have declined to 6.41 tcf as of yearend 2017 from 12.7 tcf in 2000, while crude oil reserves have fallen to 156.4 million barrels from 272 million barrels over the same period. Thailand has also some 2 Gt of lignite reserves located in the north, and there are currently limited or no reserves of hard coal. It is estimated that 1,400 Mt of lignite reserves are economically recoverable (IEA 2015).

Thailand is a major energy producer and importer. According to the IEA (2016), 36.3% of all primary energy production was gas, 31.5% biofuels and waste, 24.8% oil, and 6.5% coal. Alternative energy sources, including hydro, solar, and wind, take up less than 0.8% combined. Over the past decade, the growth in production occurred primarily in the oil and gas sector, whereas the production of coal declined. There is increasing concern over the potential depletion of oil and gas reserves over the next 8–12 years, unless new reserves are discovered to accompany Thailand's growth in domestic oil and natural gas demand. This renders Thailand highly dependent on energy imports, importing a (net) total of 65.8 Mtoe of energy in 2016, consisting of crude oil (64.7%), coal (21.5%), natural gas (17.6%), oil products (−6.5%; i.e. a net exporter), and electricity (2.4%) (IEA 2018).

Energy subsidies

In common with many other Asian economies, Thailand has a variety of subsidies in the energy and electricity sectors. The ADB notes that the largest number of subsidies was given in the form of tax breaks for diesel and market price support for LPG and natural gas vehicles (ADB 2013). Of relevance for the electricity sector is that Thailand subsidizes the price of natural gas for domestic gas consumers. However, there is an information gap about the exact differences in domestic price levels compared to the world market price. Besides that, tax incentives can be found for machinery imports used for oil and gas exploration, processing, and transport.

There are two major subsidies present in the Thai electricity sector. Firstly, electricity is supplied free of charge to low-income households that consume below a pre-determined threshold (50 kWh per month). This policy is financed by a cross-subsidy from other electricity consumers (predominantly from the industry sector). The second subsidy is provided to Electricity Generating Authority of Thailand (EGAT) in the form of subsidizing prices of lignite and natural gas for power generation. The total amount spent on fuel subsidies is unclear, owing to the lack of information disclosure about contract details. Receiving fuel at subsidized rates, however, puts an obligation on EGAT to maintain electricity tariffs for consumers at a certain level. If EGAT wants to change the tariff structure, perhaps due to changes in fossil fuel prices, it must first seek approval from the regulatory agency. The regulator makes the ultimate decision with regard to potential tariff revisions by assessing EGAT's costs of production, among other factors. Besides having its own generation capacity, EGAT procures electricity from IPPs at the market rate and distributes it to consumers at the approved tariffs. In such circumstances, the interaction between the tariff subsidy and the cost of purchasing power from IPPs determines the financial situation of EGAT and its ability to invest in new projects and ventures.

The renewable energy sector in Thailand receives subsidies in the form of feed-in-tariffs. The rates vary depending on fuel type (solar PV, wind, waste, or biomass) and the size of installed projects (ADB 2015).

Electricity sector outlook

Like most ASEAN countries, Thailand is expecting high future growth in energy and electricity demand. Currently, Thailand has the second-largest energy and electricity demand in the region. Electricity demand, in common with the trend in total primary energy supply, more than doubled between 2000 and 2017, reaching 201 GWh (EPPO 2018). The main consumers of electricity in 2017 were the industry (47%), commercial (25%), and residential (24%) sectors. Electricity access is almost universal throughout the country.

The main components of the electricity generation fuel mix in 2017 consisted of natural gas (60%), coal/lignite (18%), and electricity imports (12%). Thailand imports power from projects in Lao PDR and Myanmar, named in the previous chapters. These projects are directly connected to the Thai grid and are generally isolated from their domestic markets. The power is supplied under PPAs using a bilateral contractual form. The total installed generation capacity is about 34.5 GW, some of which is located in neighbouring Lao PDR. Over the past 30 years, the fuel mix has become more inclined towards gas and coal, while the share of oil has decreased, reflecting a worldwide trend.

The recent Power Development Plan, released by EGAT in 2015, anticipates further growth of domestic electricity demand. The Power Development Plan takes into account energy efficiency and alternative energy measures outlined in other policy directives. According to the document, electricity demand would grow 2.67% annually from 2014 to 2036. In 2036, expected electricity demand would be 326 GWh, or 80% above that of 2014.

In response to depleting domestic oil and gas resources, Thailand is likely to diversify energy supply sources through imports, coal, nuclear, and renewable energy. Fuel diversification is also one of the main topics in the new Power Development Plan. The document foresees an expansion of the domestic installed capacity from 34.7 GW to 70 GW, out of which the main additions will be renewables (21.5 GW) and natural gas (17.5 GW). Approximately 11 GW of capacity would be provided by electricity imports from neighbouring countries to Thailand (EPPO 2018).

Summarizing future developments of the Thailand's electricity sector, we conclude the following:

- **The role of natural gas will decrease but remain important.** Heavy reliance on natural gas for electricity generation and the lack of domestic gas resources create potential threats for Thailand. Diversification of the fuel mix is one of the main goals outlined in official policy documents. Even if diversification attempts are successful, the role of natural gas will still be important, as outlined in the official policy documents and projections by ADB. This could create a potential conflict of interest between subsidizing fuel for electricity and growing gas import requirements.

- **The main capacity additions will come from renewables, including solar and biomass.**
  Thailand has already taken a leading role in the development of renewable energy in the region. Although the largest renewable energy capacity is currently hydro, hydro resources in Thailand are almost fully exploited. The main capacity additions will thus come from solar power and biomass, both of which are already growing rapidly. Thailand has both abundant sunlight and a developed agricultural sector capable of providing fuels for biomass facilities.
- **The role of electricity imports will increase.**
  About 2.4 GW of power generation capacity supplying power to Thailand is located in neighbouring countries. Another 11 GW of mainly renewable (hydro) capacity is scheduled to be added over the next 20 years. Electricity imports permit Thailand to substitute the equivalent share of natural gas imports with electricity.

Development and organization of the electricity sector

Thailand's electricity market follows a single-buyer model with a liberalized generation sector and monopolies in transmission and distribution. EGAT owns and operates a significant proportion of the generation fleet, the whole transmission network, and a portion of the distribution sector.

EGAT also purchases electricity from IPPs, small power producers (SPPs), and imports from other countries (which are generally structured as IPPs). Overall, EGAT owns approximately 45% of the generation fleet in Thailand, while IPPs own approximately 38% (excluding imports). In addition, EGAT holds shares in the country's two largest IPPs, which are in charge of a 25% share in country-wide power supply.

In addition to owning and operating its own generation fleet and transmission network, EGAT is responsible for system operations, including dispatch of the generating fleet. In order to ensure that IPPs are dispatched equally with EGAT-owned generation, the activities of the market operator are ring-fenced from the rest of the company. At the same time, legislation does not provide an adequate mechanism for the market regulator to prevent the misuse of EGAT's dispatch monopoly.

The distribution sector is shared by EGAT and two other companies, Municipal Electricity Agency (MEA) and Provincial Electricity Agency

(PEA). MEA is responsible for the distribution of electricity in the metropolitan areas of Bangkok, Nonthaburi, and Samut Prakarn provinces, which together account for two-thirds of Thailand's electricity demand. MEA is not involved in power generation and has to purchase power either from EGAT or directly from Very Small Power Producers (VSPPs). PEA is responsible for generation, procurement, distribution, and sale of electricity at the provincial level, outside of the municipal zones (IEA 2016). PEA does not own or control any of the high-voltage lines within its service territory. Real-time coordination between EGAT, MEA, and PEA is managed through various regional dispatch control centres, as well as a single national control centre. EGAT also has an almost exclusive right of supplying power to MEA and PEA.

While large customers may purchase power directly from EGAT, smaller commercial and residential consumers purchase power from MEA and PEA. SPPs can also choose to supply electricity directly to consumers. IPP rights have been awarded to both international and domestic developers. Foreign investors come mainly from Asia and include J-Power (Japan), SPC Power Corporation (the Philippines), China Light and Power (Hong Kong, China), Mitsubishi (Japan), Tokyo Electric Power (Japan), Marubeni (Japan), and ENGIE (France).

VSPPs are mainly local companies that sell electricity directly to MEA and PEA. A lot of VSPPs generate renewable energy and are considered to be non-dispatchable, which means that they do not have to follow the dispatch schedule set by EGAT. Instead, PEA and MEA provide data on VSPP production to the Energy Regulatory Commission (ERC), which in turn passes that information on to EGAT.

The regulatory framework of the Thai electricity and gas sectors is formally defined by the Energy Industry Act. Energy development plans for each energy sub-sector come under one set of documents, the Energy Master Plan. It includes the Power Development Plan, the Energy Efficiency Plan, the Alternative Energy Development Plan, the Gas Plan, and the Oil Plan. Of these, the Power Development Plan, the Energy Efficiency Plan, and the Alternative Energy Development Plan are also key components of Thailand's Intended Nationally Determined Contribution. As highlighted before, the strategic direction on power sector development and interconnection is contained in the Power Development Plan.

Policies for both electricity and gas sectors are developed in Thailand by the National Energy Policy Council with the Ministry of Energy (MoE) as a key participant. The MoE is responsible for overseeing the activities of EGAT and setting the proportion of the country-wide capacity owned by EGAT. Three departments and two offices within the MoE are responsible for actual policy design and energy development plans.

The regulator functions are assumed by the ERC, established in 2008 under the Energy Industry Act of 2007. Although it operates separately from the MoE and other government departments, the ERC cannot be considered as a fully independent regulatory authority because the MoE retains some key authorities over the ERC (IEA 2016). All electricity-related investment decisions must also be approved by the Ministry of Finance.

Singapore

Country profile

The Republic of Singapore is a former British colony which merged with the Federation of Malaya for two years in the early 1960s, before it was expelled and granted independence. Singapore is a city state, located on one main island and surrounded by many small islets. Land reclamations have increased the country's total land area from 581.5km^2 in the 1960s to 721.5km^2 at year-end 2017.

Singapore has a developed, albeit small, economy. Its population of 5.6 million has one of the highest GDP per capita in the world. The economy initially grew as a refining hub for Indonesian crude oil and a petrochemical trading hub. Complemented by the country's advantageous geographic position in the Strait of Malacca, Singapore became a hub for Middle East oil exports and goods going to and from Asia.

Singapore does not have any domestic energy resources except for some solar energy potential. Its electricity sector relies almost solely on natural gas imports. Prior to 2013, Singapore obtained their natural gas supplies through pipeline connections from Indonesia and Malaysia. Since the opening of an LNG terminal, a share of the Singapore's imports is now in the form of LNG. The uptake of the LNG sector has been stimulated by imposed controls on pipeline gas imports for all new generation capacity. This policy is due to a revision in 2018.

Electricity tariffs

From 1 November 2018, the Singapore retail market became fully contestable. Prior to this date, bands of electricity tariffs applied for different consumer types. In general, these tariffs were cost-reflective and varied based on the price fluctuations of natural gas imports. Contestable consumers (typically industry and commercial customers) could either buy electricity from the liberalized power exchange, or through a retailer whose rates would be reflective of market prices. Non-contestable consumers (typically residential customers) paid a quarterly adjusted fixed tariff based on the average quarterly forward fuel oil price.

Singapore does not have energy subsidies; however it provides vouchers for population groups with low income which can be used to offset utility bills. Compared to subsidizing the cost of electricity directly, this approach presents all consumers with the true cost of electricity, incentivizing them to save and be efficient. It also does not compromise the cost-reflective pricing of electricity by the generation companies. The electricity access rate is universal.

Electricity sector outlook

Singapore consumed a 48.6 TWh of electricity in 2016. The annual growth of about 2.4% is largely due to increase in electricity in the industrial sector, which accounted for about 24% of total electricity consumption in that year. The commercial and services sector contributes to 37% of total electricity use while the residential sector makes up the remaining 15% (EMA 2017). Future electricity demand projections are not published in the official statistics; however, it is unlikely that demand will grow significantly over the next five years.

Singapore's total licensed installed generating capacity was 13.348 GW in 2017. This is almost double the peak electricity demand, which is 6.8 GW, and well above the indicative reserve margin of 30%. This capacity consists almost entirely of natural gas combined-cycle turbines. The share of natural gas in the total fuel mix was around 95% in 2017. Waste combustion and petroleum products, mainly in the form of diesel and fuel oil, and a coal-biomass project contributed the remaining 5%. The total installed capacity of solar was 99.9 MW in 2017. Singapore does not provide feed-in-tariffs or equivalent subsidy schemes for solar (EMA 2017).

Unlike most ASEAN states (with the exception of the Philippines), Singapore has a liberalized power sector with separated generation and transmission sectors. The generation sector is an oligopoly of private sector companies, with three large players occupying 75% of the market. The transmission sector is owned by the government through SP Power Grid, which manages and operates transmission and distribution networks. In order to avoid any conflict of interest, the operations of the high-voltage network are delegated to the Energy Market Authority (EMA), while SP Power Assets is responsible for the operational aspects (operating and maintaining) of the transmission system. Furthermore, given the monopolistic nature of the national transmission system, SP Power Assets is subject to price regulation.

Summarizing the details outlined earlier, we present three major factors for the future development of the electricity sector in Singapore:

- **The role of natural gas will stay dominant.**
  Due to Singapore's position as an LNG hub, and the large domestic gas-fired capacity base which has been developed, natural gas will remain the main component of Singapore's fuel mix for at least another decade. Given the lack of other viable options, gas will maintain its dominant position.
- **Renewables will continue to be a small market player.**
  Given the country's large energy consumption relative to its tiny land area, solar and other renewables will be limited to mitigating and reducing peak loads, rather than providing stable baseload power for residential, commercial, and industrial users.
- **Energy security plays a crucial role for Singapore.**
  The over-reliance on natural gas, even though it may be sourced from diverse channels, makes energy security a strategic concern for Singapore. The country has few strategic reserves to deal with interruptions to gas supply, the bulk of which is delivered by pipeline from Indonesia and Malaysia. Electricity imports can potentially help to address this problem, especially if they come at a cheaper cost than domestic supply. However, for electricity imports to be viable, Singapore has first to deal with conflicting interests within the domestic power industry.

Development and organization of the electricity sector

The electricity sector in Singapore is regulated by the EMA, a statutory board under the Ministry of Trade and Industry (MTI). The EMA also regulates the natural gas market, ensuring a stable source of fuel inputs for the sector. In the electricity sector, the EMA is responsible for approving technical and market standards, issuing market licenses, and protecting consumer interests. Energy policy is set by the Energy Division of the MTI, with the goal of ensuring energy security while maintaining economic competitiveness and environmental sustainability to support long-term economic growth in the country (EMA 2017).

Singapore's electricity market

The National Electricity Market of Singapore is a power exchange similar to other competitive electricity markets described in this book. The associated financial market is run by the Singapore Exchange.

The power exchange uses an auction pricing arrangement, whereby generators bid in their price-quantity stack for every half-hour, which is matched to the load forecasts produced by the market operator, EMC, for each bidding period. The resultant market clearing price, which equates demand and supply, is the marginal spot price in the wholesale market. Similar operations are conducted in the ancillary services market, which determines the reserve and regulation price and capacity for each generator.

The market is cleared for a system of nodes across the island which provide a better resolution for congestion management compared to zonal pricing, although the price differential between different nodes is very small in Singapore.

Full liberalization of the energy market occurred on 1 November 2018, when the retail sector became fully contestable. All consumers are now free to select and purchase power from their preferred retailer rather than from the former monopoly retailer, SP Services.

References

ADB (Asian Development Bank). 2013. *Lao People's Democratic Republic: Energy Sector Assessment, Strategy and Roadmap*. Philippines: ADB. Accessible at: www.adb.org/sites/default/files/institutional-document/33914/files/lao-pdr-energy-assessment-2013-update.pdf.

ADB (Asian Development Bank). 2015. *Fossil Fuel Subsidies in Thailand. Trends, Impacts and Reforms*. Accessible at: https://www.adb.org/publications/fossil-fuel-subsidies-thailand-trends-impacts-reforms.

EIA (Energy Information Administration). 2017. *Country Analysis Brief: Malaysia*. Washington, DC: United States Energy Information Administration. Accessible at: www.ieee.es/Galerias/fichero/OtrasPublicaciones/Internacional/2017/EIA_Malasya_26abr2017.pdf.

EMA (Energy Market Authority). 2017. *Singapore Energy Statistics 2017*. Singapore: EMA. Accessible at: www.ema.gov.sg/cmsmedia/Publications_and_Statistics/Publications/SES18/Publication_Singapore_Energy_Statistics_2018.pdf.

Energy Commission of Malaysia. 2016. Peninsular Malaysia Electricity Supply Industry Outlook 2016. Accessible at:https://www.st.gov.my/ms/contents/. . ./outlook/Outlook%20PM%202016.pdf.

EPPO (Energy Policy and Planning Office). 2018. *Electricity Statistics*. Thailand: Ministry of Energy. Accessible at: www.eppo.go.th/index.php/en/en-energystatistics/electricity-statistic?orders[publishUp]=publishUp&issearch=1.

ERIA (Energy Research Institute for ASEAN and East Asia). 2015. *Study on Effective Power Infrastructure Investment through Power Grid Interconnections in East Asia*, Accessible at: http://www.eria.org/publications/study-on-effective-power-infrastructure-investment-through-power-grid-interconnections-in-east-asia/.

IEA (International Energy Agency). 2015. *Development Prospects of the ASEAN Power Sector*. Paris: IEA.

IEA (International Energy Agency). 2016. *Thailand's Energy Security Assessment 2016*. Paris: IEA. Accessible at: www.iea.org/publications/freepublications/publication/Partner_Country_Series_Thailand_Electricity_Security_2016_.pdf.

IEA (International Energy Agency). 2018. *World Energy Balances*. Paris: IEA.

Ngadiron, Z., N. H. Radzi, Z. Yassin, and I. Amin. 2014. *Review on Restructuring of Malaysia Electricity Supply Industry*. Working Paper. Accessible at: http://eprints.uthm.edu.my/6406/.

Ulrich, Z. 2015. *Can We Make Hydropower Work for All in Laos?* Accessible at: www.worldbank.org/en/news/opinion/2015/05/14/can-we-make-hydropower-work-for-all-in-laos.

UN (United Nations). 2017. *National Energy Data Lao PDR*. Accessible at: https://unstats.un.org/unsd/energy/meetings/2017a/5.2meml.pdf.

World Bank. 2016a. *Lao PDR: Overview*. Accessible at: www.worldbank.org/en/country/lao/overview.

World Bank. 2016b. *Thailand: Overview*. Accessible at: www.worldbank.org/en/country/thailand/overview.

World Bank. 2017. *Access to Electricity (% of population)*. Accessible at: https://data.worldbank.org/indicator/EG.ELC.ACCS.ZS.

Index

Note: page numbers in *italics* indicate figures, in **bold** indicate tables.